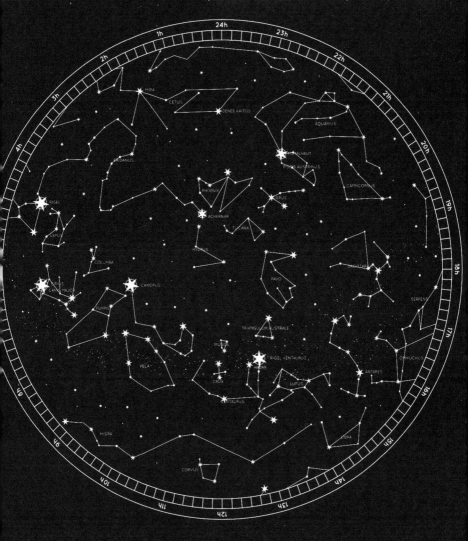

SOUTHERN
HEMISPHERE

The Secret World of

STARGAZING

The Secret World of

STARGAZING

Find solace in
the stars

VIRTUALASTRO
ADRIAN WEST

For Karen, my children and my parents.

First published in Great Britain in 2021 by Yellow Kite
An imprint of Hodder & Stoughton
An Hachette UK company

1

A CIP catalogue record for this title is available from the British Library.

Hardback ISBN 978 1 529 38207 5
Ebook ISBN 978 1 529 38208 2

Typeset in Garamond by Goldust Design
Illustrations by Goldust Design
Printed and bound in Great Britain by Clays Ltd, Elcograf S.p.A.

Hodder & Stoughton policy is to use papers that are natural, renewable
and recyclable products and made from wood grown in sustainable
forests. The logging and manufacturing processes are expected to
conform to the environmental regulations of the country of origin.

Yellow Kite
Hodder & Stoughton Ltd
Carmelite House
50 Victoria Embankment
London EC4Y 0DZ

www.yellowkitebooks.co.uk

CONTENTS

INTRODUCTION

Do you look up at the night sky and wonder? If you do, then this book is for you, and every member of your family, young and old. It's for your friends, aunts and uncles, parents and grandchildren, neighbours and colleagues. Even your cat or dog – if they could read, that is.

There are billions of humans on this planet, and only a tiny fraction of us understand and enjoy the night sky, or even remember it's there sometimes. I wrote this book to enable and encourage more people to look up and understand as much or as little as we want to about what can be seen in the dark night above. I want everyone to delight in this aspect of the natural world a little more, and to feel the benefits to our health and wellbeing of picking up the stargazing habit.

Starting young

I started my own stargazing journey at a very early age really, as a toddler in a rural Oxfordshire village with amazing dark skies. My grandparents lived just two or three doors away from my home. Because of that, my parents and grandparents would go from house to house most days. We'd often have dinner later in the evening at their house, and when we'd return home the sky would be filled with stars on a clear night. Luckily, our small village had no streetlights.

I was amazed by the night sky and nature in general, and my parents and grandparents would encourage this, making up stories about the stars for me. They weren't astronomers themselves, but they did their best to nurture my enthusiasm. I also remember lots of travelling to visit family as a child, and I would be stuck in the back of the car really late at night going home and just gazing out the window, watching Orion up above. I think I witnessed my first shooting stars on a trip like that.

As I grew up, I was always drawn to science fiction in books and magazines, and on the television. I think the key moment for me was actually *Star Wars* in 1977, when I was about six years old. My mind was totally blown away because I could look up into the night sky and to me, this stuff was going on for real up there, out there.

It was these thoughts that made me feel wow, this is brilliant. I enjoyed a lot of the TV programmes, such as *Tomorrow's World*, *The Sky at Night* and *Space: 1999*. Astronomy came from my interest in both science fiction and science. Growing up in the 70s and 80s, every science fiction TV programme would tell you that by 2021 we would be flying around the universe.

And these things fired me up – I was looking to the future. As a child, the future was bright according to programmes like *Tomorrow's World*. But as I got older, I became more realistic about science. I was interested in technology, but also natural science – geology, the world at large, and the universe, space and astronomy.

As an adult, I became involved with astronomy groups. I have written articles for well-known online astronomy magazines and have provided content for the web and TV. I do group guided Night Sky Tours for prominent organisations such as the National Trust, and many other stargazing related activities. I have run a popular website for the past ten years and have a massive following on my VirtualAstro channels on social media.

Astronomy for real people

It was actually when I met one of the key presenters of *Tomorrow's World*, Maggie Philbin, who got me to where I am. About 12 years ago, I met Maggie at an International Year of Astronomy event. This was one of my childhood idols, talking to me. Once I stopped being starstruck, we had a great discussion and she introduced me to Twitter. My Twitter journey started as I wanted to encourage a new way of looking at science and astronomy. I thought that the traditional astronomy scene could be quite stuffy and unappealing for the general public. Astronomy events felt awkward and formal, and I loved talking to normal people who didn't know anything, and who would look up and ask the obvious questions: 'Where do I see this? What time can I do that?' That is why you often find me doing guided tours to groups on the side of a windswept hill, on a theatre stage or online – talking stars with normal people who just want to know more.

My brain is not mathematically built like professional astronomers. My interests are in a natural form of astronomy rather than the depths of astrophysics. I think my major passion is the actual beauty of the night sky, and what is to be seen in the motion of the sky as it changes with the months. And I see that this is what people are generally most interested in.

Astronomy is a fascinating and rewarding activity but people often tell me that they think it involves a lot of maths and physics and that it's 'not for them', that it's inaccessible for those with just a casual interest, or who aren't scientifically or technically minded. I'm here to prove you wrong. We won't be doing any of that stuff in this book – we don't need it to be able to find and view an astonishing array of celestial offerings.

No equipment required

Historically, astronomy has always been viewed as rather elitist: the domain of academics, scholars, school kids and retired old men with lots of time and money – construed as something not for everyday people. The night sky can certainly be confusing. There is lots to see, and much of it looks chaotic and random. But I want stargazing to be accessible to everyone in the way that, in a sense, it always has been – with the naked eye.

Many of us just enjoy looking up at the night sky and taking it all in. Some look up in the dark to wish, to dream, to get creative or simply to relax. There are countless emotions and thoughts, all different and personal to each stargazer, but there are so many, many questions.

So I have written this book to answer many of these questions and to show you that you don't have to be a scientist or a grade A student at school to enjoy, navigate and understand the night sky. Stargazing isn't hard, and I want it to be enjoyable and fun.

The reward for spending a few hours reading *The Secret World of Stargazing* is that you will be able to navigate your way around the night sky, and in doing so will find yourself more in tune with the seasons, the planet and the nature around you.

It's not just about stargazing; it's also about you. Being outside at night taking in the sights of the universe is an awe-inspiring, invigorating, uplifting and soothing experience – unique for everyone. This is an opportunity to enrich yourselves by doing something wholesome, interesting, fun and good for you and for the planet, with your whole family or friends.

Together, we will learn how to understand the night sky's motion and movement, to identify some of the most prominent stars, and learn the more familiar constellations (the shapes in the sky) to help you navigate your way around the sky using these as signposts.

I'll tell you when and where to look for planets, meteor showers, the International Space Station and other things that mysteriously appear and move in the sky, all in an easy-to-understand and non-technical manner. *The Secret World of Stargazing* is not an academic textbook, nor is it technical or heavy-going. No prior knowledge is needed.

Astronomy is a rich, varied and fascinating subject that you can study at your own pace. With this book, you'll become a competent stargazer, and this is an excellent stepping-stone into the world of astronomy and scientific study if you want to learn more. You'll also learn to relax and quiet your mind as you stargaze.

So, let's take our first steps to enjoy and understand the stars together.

Chapter One

STARGAZING IS GOOD FOR YOU

When I get onto my soapbox about how stargazing is one of the simplest and most magical activities you can embark on to strengthen your mental wellbeing, some people ask, 'How can that be? Don't you just look up?'

But there's more than that to watching the stars in so many ways. The sounds, the smells and the very act of standing, sitting or lying outside and stargazing are refreshing and rewarding. In this chapter, I want to share one of stargazing's biggest secrets: how it can make us all feel better, happier and more at ease with ourselves.

Wellbeing is one of those buzz words you hear a lot nowadays. More and more of us seek out rich and varied ways of being healthier, living longer and feeling happier. Common ways to do this might involve taking up a sport, exercising outdoors or going to the gym. We try to get out in nature more – especially in the warmer months – with walking, jogging and open swimming popular pastimes. Often, we look for the means to bring stillness and calm into our over-busy and demanding modern world, to carve out time for ourselves and let our minds unwind in a good book, hobby or even computer game. The list of possibilities is endless . . . but you rarely see astronomy or stargazing on the list. Why is that?

Just stop and stare

Stargazing is the part of astronomy we can all enjoy. I think that the main reason astronomy is overlooked as an enriching and healthy outdoor pastime is because it is traditionally seen as a science – and educational, so is largely disregarded as a health and wellbeing opportunity. But stargazing, once you strip away the heavy-going astronomy, science and technical bits that can turn people off, is refreshingly fun and so very easy.

One of the things I love most, though, is that you need nothing to get started. Of course, many do study it professionally and buy extensive equipment to help develop their interest (I talk about telescopes and other paraphernalia later in the book). There is no requirement to buy equipment though, to find time in a hectic schedule, buy a subscription, arrange childcare or a pal to go with, or wear a certain outfit (though warm clothing will often come in handy). You can bring a comfy chair to sit on but a patch of grass to sit or lie on is just fine too – or some people lie on their trampoline. You can take or invest as little or as much time as you like, on whatever night of the week fits into your life. You don't need any prior knowledge or skills. Your only requirement is a clear sky. It's just you with nature. That's it.

The natural world/nature

Birdwatching is a popular hobby that encourages people to venture out in nature, to explore the world around us, take in lots of fresh air and enjoy a relaxing experience. Similarly, gardening is a common pastime in the great outdoors, inspiring us to learn about plants and the earth (and studies show the broad health benefits of getting our hands into the soil). Why don't we associate those same benefits with stargazing?

The night sky is an integral part of the natural world. Just as the land around us reflects the changing of the seasons, the circle of life and the passing of time, so too do the stars above.

It's yet another way to connect with Mother Nature, plus it also involves the magic of darkness.

Night-time is something special – different smells, different sounds and a new view of everyday surroundings. As our senses sharpen, the everyday demands of life diminish. For many, it is the time when children sleep, emails stop pinging and we are able to disconnect from our worldly problems. Finally, we have time to stop and look. To be present in our environment, and to explore what time outdoors means to us.

Life in perspective

I call stargazing a 'band aid for the soul': it has the same effect as sitting on a beach and staring at the ocean's waves, or walking through a deep green forest. The environment washes over you and it alters your state of mind, as well as calming your breathing and heart rate – helping you to relax and to relieve tension. It brings clarity, and knocks back the noise and the anxiety; the day's stress is washed away in the starlight as you look up and let your mind wander.

There's a passion for 'slow living' at present and stargazing represents that desire to move away from our 'everything now' culture, to increase our awareness of the here and now as we observe the skies, searching out constellations. Looking up fills us with unadulterated awe – we have no negative associations or fears of the night sky, and following the stars and learning about them gives us a clear sense of our permanency in the world, and our sense of place. Frustrations, grievances and struggles are pushed further away as we immerse ourselves in the immediacy of the night sky.

I notice the effects when I do my Sky Tours with the National Trust. I have up to 40 people – adults and children – at the event, and at the beginning everyone's a little anxious. There are questions about the night and people are waving their torches around; children are running around and being noisy. But I've noticed that, after a while, everyone relaxes. They ask deeper questions. Even the children are calmer, looking up and enjoying themselves, not running around.

Stargazing doesn't just help your mental health either. Ongoing stress and poor mental health are damaging for your physical body, and can contribute to health issues such as high blood pressure, heart problems, diabetes and other illnesses. Like meditation and mindfulness, I want to prescribe half an hour of stargazing every single night.

Some have told me that stargazing makes them feel like they do when they stroke a kitten. Others have compared it to relaxing in a hot bath, but without the hot water and tub. (But if you want to lie in a bathtub and look at the stars in your garden, go ahead.) It draws you into the present moment like a meditation/mindfulness session, as you embrace the dark, the night environment and zone out from daily life.

Recent times

I have heard many share their stargazing experiences and how much it has helped their health and wellbeing, both via social media and face-to-face. From people reminiscing about fond memories of camping under the stars or lying on a beach on holiday and observing the night sky, to those walking home from a night out who ended up spending hours gazing upwards as something above grabbed their attention. All expressed immense feelings of peace, awe, contentment and curiosity.

The benefits for our mental health became overwhelmingly evident while we were all in lockdown during the Coronavirus pandemic. It was a tough time. The health and wellbeing of a lot of people were impacted.

As most places were out of bounds and closed, and mixing outside of your own household was also restricted, many families were forced to find alternative ways to entertain themselves. It was a hot spring and summer for many, so people ventured outside into their gardens, back yards and any outdoor space they could find.

And something happened, something profound. On those warm summer evenings, people started to look up! The Sun would set, and the stars would come out. The sky's pollution had cleared with much less traffic on the roads than usual, and the universe presented itself to everyone in all its beauty and splendour.

People from all walks of life were captivated for the first

time. It was a seismic shift for astronomy and stargazing. I don't believe there have been so many people looking up at the night sky in recent times. As a result of this new perspective, and the opportunity to explore the heavens above, many saw the Milky Way stretched across the sky for the first time and faint streams of newly launched internet satellites traversing the darkness like a celestial string of pearls. They tried to distinguish the brightest stars in the sky and looked for planets visible with the naked eye, while noting that what appeared above them moved as the seasons changed, and that stars and celestial objects came in varying shades and brightness. The International Space Station mesmerised as it passed over, many waving to the astronauts on board. The planets shone, and shooting stars sparkled. We saw new objects for the first time; there was even a rise in UFO sightings and reports.

Due to this almost overnight mass interest, people wanted to find out more. My own social media channels expanded hugely in less than a month. Everyone wanted to know what they were viewing with their naked eye, and what they could look for next. Questions and requests for guidance kept coming in. Lockdown made us grateful for any nature on our doorstep, and the clear skies were right there.

And with that was the sheer volume of people expressing how good the night sky and stargazing made them feel. Doctors and nurses who had been working hard in intensive care wards were telling me how soothing and mentally relaxing looking up at the night sky was for them after long shifts. Those struggling with stress and grief found themselves able to think more clearly and pull themselves out of the darkness and back into the light by looking at the stars.

Families were able to bond over this new activity, it engaged all ages and generations and the lack of skills and learning needed meant that no one was excluded. A friend of mine started stargazing with their teenager during exam time. They found that time sitting in darkness, in near silence, together for half an hour each night helped hugely to settle both their whirring minds and levels of stress.

I, like many other people, enjoy many activities that help make me calmer, healthier and happier, but stargazing is the big one that ticks all the boxes. It's at the top of my must-haves for my mental health. I feel the benefits as soon as I relax into the night sky.

Sparking creativity

People also tell me how watching the sky boosts their creativity and helps them with their thoughts. Since humans have looked up at the night sky, it has sparked imagination and innovation, religion, literature and art – from the early hunter gatherers spreading across the world through to the earliest civilisations, such as the Mesopotamians (located in modern day Iraq). Depictions of constellations such as Orion have been found on 30,000-year-old mammoth tusks, and one of the most famous astronomical landmarks – Stonehenge – was built over 5,000 years ago. Indeed, astronomy is the oldest of the natural sciences and was developed to an advanced level by the ancient Greeks. We still use much of what they understood and mapped today. Those who explored and expanded our understanding later are household names: Copernicus, Galileo, Kepler and Newton.

The night sky has always fired our thought processes – and the calming views help me open my mind and concentrate. Sometimes you don't even have to focus, ideas or thoughts just present themselves. It makes me think of famous writers and creatives in history who would take themselves off on retreat to shut out the rest of the world and use the tranquil environment to allow their natural creativity the space and time it needed. The night sky has the same function.

Learning about the stars at your own pace is rewarding also. Learning new things or finding out about stuff we have an interest in fills us with good feelings. Trying new experiences and pushing our brains into new habits is invigorating, challenging and, from that, comes the gratifying sense of pay-off as we gain new skills and knowledge.

You're in charge

You can, of course, take stargazing as lightly or as seriously as you wish. Some may want to see a few well-known constellations or locate a particular star, or just want to take it all in and enjoy the view. Others will want to hone their skills and knowledge by venturing deeper into the world of astronomy and beyond. You may think about investing in some equipment. It's entirely up to you. Those who haven't done it before will tell you how they soon become hooked and wanted to learn more. However you decide to stargaze, it's such a fascinating and accessible hobby with which to energise and stretch your brain.

Now that I've convinced you what a difference exploring the night sky can make to your life, it is time for us to take our first steps to understanding and enjoying the stars and the wonders within, learning how to enjoy and understand their scale and movement. We will tune in to the sights, sounds and smells of the night and immerse ourselves into the natural world and beyond.

Remember that the key is to enjoy yourself; there's no test at the end. See! It's making you feel more relaxed already, so let us begin.

Chapter Two

START LOOKING UP – THE SECRET WORLD OF STARGAZING

Before we jump in and start to explore the night sky, this is an excellent time to ensure that we understand the motion and movement in the heavens above, and the natural, or even un-natural, world around us.

As I re-enforce throughout this book, stargazing does not have to be complicated. It can be as technical and challenging as you want it to be. Astronomy, physics and science in general are extremely rewarding and it would be fantastic to get more people interested, but for stargazing, we can understand and enjoy the universe in less precise or completely different ways.

The map in the sky

Since the dawn of mankind, humans have used the stars and their changing positions, and animals have been doing the same for a lot longer. Many animals – mammals, birds, fish and insects – are aware of the night sky and some have evolved to a point where their behaviour is in part controlled or influenced by it. They understand the world of stars in their own way; to them it has meaning. We aren't just talking day or night, if animals are nocturnal. Some animals use the night sky to help them to find food, hunt, navigate and migrate under the cover of darkness. Many – such as certain migrating birds – use the night sky for visual cues and landmarks as other animals would do during the daytime.

It's easy to gloss over their natural knowledge and ability, which is way beyond ours. For example, we're still not sure how some of the migratory bird species make their epic journeys each year. Many prefer to migrate at night and use the stars to navigate. We only know that they use the North Star – also known as the Pole Star or Polaris – when they fly south, and that they have a type of internal magnet within them. The North Star is a fixed point in the sky – which I'll explain more about shortly – the brightest star of its constellation, Ursa Minor (the Little Bear), and extremely close to the position of the North Pole. The sky moves around it, as it were. The birds, it is believed, use this stationary point in the sky in order to travel south, regularly checking that the North Star is behind them.

Other marine and land species use bright stars and the shapes they make in the different seasons as short-term markers for navigation. Seals hunting for food in the sea at night don't have identifiable terrestrial landmarks but can identify bright stars and the shapes they make, which they use to navigate and orientate themselves. Ancient sailors did the same, of course, and these stars were known as lodestars in the past. But the seals have been using them to navigate for much longer.

Unlike birds and mammals, insects have what are known as compound eyes, and are unable to see detail and points of light such as stars, as we do. However, an insect's eyes are capable of seeing the light of the Moon as a point of reference, and on a moonless night they can also see the river of stars that form the Milky Way as a bright band of light stretching across the sky. They use this strand of light to navigate in a particular direction or in straight lines. (Later in the book, I'll explain how you can see the Milky Way.)

So, how does this help us when we are stargazing? I wanted you to understand how animals, insects and ancient people used the sky for a straightforward thing, navigation – signposts in the sky – so that you can appreciate that simplicity is one of the secrets of stargazing. The night sky contains signposts from one star or shape (constellation) to other shapes, that then lead to other objects. Think of it as a giant map. I like to tell people that if you can find your way from A to B or around a supermarket, then you can stargaze!

Before using any map, though, it's always good to have an idea of where you should start and where you want to go. Your surroundings are as significant as what you are doing when you stargaze, as with any journey . . .

How to watch the stars

There are two kinds of stargazer: accidental stargazers, and those who want to take some regular time to stargaze or do astronomy.

The accidental stargazer is someone who hadn't planned to look up. You could be getting out of the car after a night-time journey, walking home from a night out or taking the dog for a walk. Then the night sky catches your eye, and you look up and contemplate it for a few moments. We've all done this, and it's a perfect time to stargaze. Maybe you pick out a prominent object or see something like a shooting star moving above, or perhaps you just notice one or two bright stars shining strong on a clear night. Then you carry on with what you were doing. There's no time limit or pressure. These random moments never last long enough for me – usually because I'm too involved in doing another task, or I could have dressed more appropriately and am freezing!

The other type of stargazing is when you have time to

search purposely. Planning ahead a little might help in this instance. Maybe the weather forecast is clear, or you have heard there is something cool to look at in the sky that night, or you need some quality time to relax and want to take in its wonders. It doesn't matter what the reason is, but to get the most out of being outside in the dark it is a good idea to prepare yourself so that you can do so for as long as possible.

The most important preparation for any night-time activity, including stargazing, is being warm and comfortable. If it's winter and -5° outside and you venture out barefooted in shorts and a T-shirt, your stargazing session is going to be brief at best. Even in the summer, a slimy slug squidging underfoot and oozing between your toes will be enough to hasten your retreat indoors.

Dress appropriately and you will enjoy it. Even in the more temperate months, it can get cold and chilly at night so it is a good idea to have some warm clothes and footwear standing by, and even a warm hat and gloves. In the colder months, this goes without saying.

Where to go?

Clothing isn't the only thing to consider for your stargazing comfort and an enjoyable session in the park or garden. You could be outside looking up for a considerable amount of time, so your stargazing spot is just as, if not more, important. This could be as simple as standing out in the street, sitting on your balcony or in your back yard. Some of us have gardens but many don't. It doesn't matter where you are, though, as long as you can sit, recline, lean or even lie down. It's about being comfortable and having a good view of the night sky for as long as possible.

If your view is terrible, try to get to a park, green space or open area where you can safely see the sky. If you live in a built-up area such as a town or city, you will probably have light pollution to contend with. All you can do is shield yourself from direct light sources such as indoor lights, security and street lighting; unfortunately, the light pollution above will drown out all but the brightest stars, but you will still see some.

The good news is that many local authorities worldwide are aware of light pollution and its effects on the health and wellbeing of humans and wildlife that live in urban areas, not to mention the cost and impact of climate change. Many now switch off street lighting late at night for a few hours and have replaced old street lighting with smarter lighting. In a town I visited recently the light pollution was awful until, suddenly, the streetlights went out at 11pm and much more of the sky became visible. Improving lighting practices in general will eventually restore or significantly reduce light pollution in built-up areas, so there is hope.

If you can't get away from light pollution, you could travel out of your urban area into the countryside or to a National Park, many of which are also dark sky reserves.

For a special evening of stargazing, you could visit Exmoor National Park and Galloway Forest Park in the UK, the Grand Canyon in Arizona, USA and Warrumbungle National Park in Australia. Most countries have dedicated dark sky reserves, not to mention dark rural countryside areas. Once you have settled comfortably into your observation spot, make sure you have everything you need for a half-hour, hour or a long evening stint. There's no time limit – look up for as long as you want.

If you are lucky enough to have a good pollution-free view of the night sky, allow a few minutes for your eyes to become accustomed to the dark. We call this dark adaption, and it will enable you to see a lot more. Without getting into the biology, your eyes can take several minutes to get used to being in the dark and become sensitive enough to see the light from the faint stars. Under no circumstances should you look at your phone or other devices or shine torchlight into your eyes while you are star spotting; this will ruin your dark adaption, and you won't see much for quite a while afterwards.

If you use a mobile device such as a smartphone specifically for stargazing, you can attach some red film available from hobby or lighting stores, or use apps with an astronomy/dark mode. Red light helps you maintain your dark-adapted sight, so you can use a rear bicycle light as a torch or buy a red-light torch. Just don't have it too bright or shine it in your eyes.

Finally, be mindful of your chosen stargazing spot as I

regularly terrify people walking by if I'm sat by the hedge in my front garden. Be courteous to your neighbours and careful where you stargaze. It helps to let people know where or if you are in a public place or the darkest depths of your garden.

Day and night

Now, imagine you are in your chosen spot; you are comfortable, dark-adapted as best you can, and ready to stargaze. One of the first things you need to understand is the sky's motion.

If you look at a particular star or object, let's say the Moon, you will notice as the minutes and hours go by that it moves and doesn't stay in the same position as where you first spotted it. You will notice that it has moved in a westerly direction. You may spot that it isn't just the object you were observing that has moved, but the whole sky.

Many people think the night sky doesn't move, and they are right in one respect – the movement is from the Earth. The Earth is rotating underneath the sky, like a giant spinning top or a ball. One complete rotation of the Earth is 24 hours or one day. This is why we experience both day and night with each rotation, with the Sun, Moon and everything else appearing (rising) in the east and disappearing (setting) in the west. Day happens when the part of the Earth on which we live is facing the Sun, and night occurs as our part of the planet is facing away from the Sun.

The North Star at the centre of rotation

I mentioned the North Star earlier in this chapter and its importance when navigating and understanding the night sky. If you can imagine pushing a pencil through a small ball of clay or Play-Doh and then using this pencil to spin the ball, you are emulating the Earth's rotation – each rotation in one day. Now visualise this as a massive pencil going through the actual Earth, from the North Pole to the South Pole – this is the Earth's 'axis', its centre of rotation. This centre of rotation appears to be fixed in space above the North Pole at the North Star, also known as the north celestial pole (and again, the Pole Star or Polaris). This is a fixed point in space as the Earth rotates while everything else moves around it. You can see this quite clearly in long exposures or star trail pictures.

The North Star or north celestial pole is only visible in the northern hemisphere, and not the south. So, what about the southern hemisphere? We mustn't forget that the same thing is happening in the south as in the north; the only difference is a fixed point of the Earth's axis isn't as easy to find or as apparent in the south. The fixed point is still there above the Earth's South Pole, and all other objects move around it as they do in the north, but it's not as simple to read in the stars. We will look at how to find both the north and south celestial poles later.

The Moon

The Earth's rotation isn't the only thing making notable moves or changes in the night sky. The Moon has the most noticeable amount of movement compared to almost any other object visible at night (meteors, shooting stars and artificial objects being the exceptions). The Moon will noticeably change in appearance and position from night to night; you can even see a slight amount of movement throughout a single evening.

Why? Remember that the Moon orbits the Earth in a big circle, taking one month (the word 'month' comes from 'Moon' and denotes what we call one lunar cycle). As the Moon moves through the month and its orbit/lunar cycle, it will appear in different parts of the sky at different times. Not only is there this movement, but its appearance changes with the amount of sunlight hitting it; these are known as phases, such as when the Moon is Full, half Full, or at First Quarter (and I'll explain in more detail later).

The planets

Along with the Moon, you may notice the odd star looking out of place. Some of these stars are very bright, outshining all other stars in the night sky; they often appear near or along the same path that the Sun and Moon take in the sky. These aren't actually stars, they are planets.

There are eight planets in our solar system – which is what we call the planets and their moons that orbit the Sun. Five of those planets are visible to the naked eye – Mercury, Venus, Mars, Jupiter and Saturn. Of the other three planets, one is our home, the Earth, and Uranus and Neptune require a telescope to be seen.

You can observe the motion of the solar system as the planets move slowly across the sky over weeks and months as they rotate around the Sun in their respective orbits, and we look at how to locate the Moon and planets later, in Chapter 10.

There are also countless space rocks, comets and dwarf planets (spherical objects not large enough to be deemed planets) within our solar system. You can see some of these just using the naked eye, and we'll learn more about these in Chapter 11.

The celestial sphere

To find our way around the night sky we treat it like a dome above us, with the stars attached. We call this the skydome, or you can think of it as the inside of a sphere – the celestial sphere. As our eyes adapt, and if we keep our gaze fixed, we begin to see further away from our planet and solar system to a sky full of stars beyond – if the sky is clear, of course.

The stars and everything in the night sky appears to

be projected onto the inside of the dome shape, with the Earth below. Some ancient civilisations did believe the Earth was flat and the sky was an enormous dome with the stars placed on it, or the light of the gods shining through holes. Modern science has proved that the night sky is infinite and the stars and objects within are at various distances from us – and these distances are vast. But the key to understanding is to remember that the night sky appears to move around the planet due to the Earth's spin.

A year of seasons

There are two very distinct forms of movement in the sky that we need to understand. As well as the daily rotation of the Earth, remember that the planet is also orbiting the Sun. This complete circle or orbit of the Sun takes one year. As the Earth moves around the Sun through the year, its position relative to the stars of the solar system and beyond therefore changes.

Try this exercise: imagine you are in a room with a lightbulb in the centre (representing the Sun). Now walk around the room circling the lightbulb. Start by facing the lightbulb and then turn slowly as you move around the lightbulb so that you make one complete 360-degree turn around the bulb before you return to the spot where you started. At first you will only see the lightbulb but as you walk around the room, turning slowly away from the

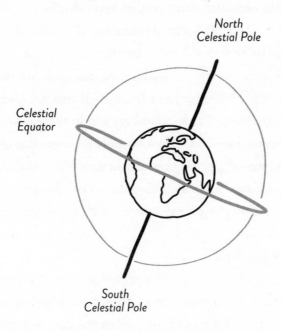

lightbulb and then back towards it, you will see all of the room's different features at different times in your route. The furniture and walls stay in the same place but they are seen from different angles as you move, and sometimes you can't see parts of the room at all. After completing one circuit or orbit (one year), you arrive back at the same position and view as when you started.

This simple exercise shows how the view of the room – the night sky – changes as you move through the months and seasons. You could also spin 365 times to emulate all the days of the year as you orbit if you wish. Make sure you have a friend recording this for when you fall over!

Of course, as we move through the year, everywhere on the planet experiences the four seasons – denoted by the hours of daylight and the average temperature. The seasons are a result of the Earth's tilt relative to its orbital plane with the Sun. As the Earth orbits over a year, the poles are either pointing toward the Sun (summer) or away from the Sun (winter) depending on which hemisphere you are in. In the summer, we have longer hours of daylight and shorter nights, and vice versa in the winter – shorter days and longer nights.

We'll learn more about the Moon, planets, stars, constellations and seasons in later chapters but for now this gives you a basic idea of the night sky's movement to help you understand more of the natural world when you stargaze.

Out at night

Finally, when you are out stargazing, there is one more thing I recommend you consider: the world around you. Stargazing isn't just about looking at the sky above; it's about being still in the dark at night – something we don't do enough. In the last chapter, we talked about the health and wellbeing benefits of stargazing and I believe 'tuning in' to the world around you at night to be one of the essential parts of stargazing, and the most rewarding. You aren't just using your eyes to look at the sky; all of your senses are heightened. Pay attention to the sights, sounds and smells of the night and immerse yourself in the natural world and beyond.

In the daytime, everything feels familiar and the cacophony of life around us numbs our senses. However, once it is dark, our familiar surroundings become un-familiar and it feels like our world has changed. The senses work harder to understand and navigate, with your sense of touch, smell and hearing becoming more acute. As your eyes adapt to the darkness, you start to pick out more detail in your surroundings, and see the faintest points of light. You are more able to smell the grass and plants around you and, in the colder months, you may pick up a waft of coal and logs burning as the smoke drifts down from people's chimneys. It's one of my favourite winter scents and conjures up fond memories. The noise of traffic and hustle and bustle have stopped at night so the smallest sounds can be heard close by, and distant noise travels with exceptional clarity.

It doesn't matter where you are on the planet. The night sky lets you be part of the goings-on of the nocturnal natural world and this is what also makes stargazing unique. Not only do you immerse yourself in the sky above, but you also gain an intimate awareness of the secrets of the night.

You are now set up for an enjoyable, relaxing and fulfilling night under the stars, comfortable and able to make the most of your observation spot. You understand the basic motion of the night sky, and the movements of the key elements within it, a vital start to making the most of stargazing. It's time to look up . . .

Chapter Three

THE NIGHT SKY

Are you comfortable and ready to stargaze? In this chapter, we will look at the night sky in more detail, with some simple tricks to help make your exploration more fruitful and enjoyable. These will help you find your way around and understand what you are looking at and, as promised, there won't be any heavy-going science or explanations.

Twinkle, twinkle

When we look up on any clear dark moonless night, we are presented with a sea of stars, like diamonds encrusted on black velvet. You may see different shades and colours of stars, with only the brightest popping out and grabbing your attention.

There are 'large' stars, faint stars and stars of all levels of brightness. When you watch for a while, you will notice that you can see white stars, yellow stars, and some that look blue or red or orange to the eye – stars of all colours. There are some exceptions such as purple and green; you won't see any of those as the other colours emitted by stars mix and hide green and purple from human eyes.

So, what are stars? One of the easiest ways to understand what stars are is by looking at the closest one to us – the Sun. Yes, the Sun is a star just like many you see in the night sky, and it provides all the energy our planet needs for life to grow and survive. Like the Sun, all of the stars in our galaxy and beyond are basically giant balls of superheated gasses. (It's easy to forget this as the tiny whitish dots look very innocent in comparison to the large burning Sun.)

If the Sun had never formed, our solar system wouldn't exist, nor would we, or this book. The Sun is the reason we are here. Our entire solar system is held together by the massive amount of gravity (the thing that keeps you and everything else on the ground and stops you floating away) from the Sun caused by the huge amount of material within it.

Our Sun (and the solar system) was born about 4.5 billion years ago from a massive cloud of gas and dust called a nebula, which collapsed in on itself under the weight of its own gravity. The cloud flattened into a disc as it spun and, at the centre, a star formed – the Sun – and sucked in most of the material from the nebula. The gasses in the Sun were compressed under tremendous pressure, fusing atoms at its core within. Stars produce their own energy and this nuclear fusion at the core powered the Sun and resulted in intense heat and light. Our solar system was then born, including eventually Earth. Indeed, everything on our planet originally came from this nebula, including you and me. We are truly made of stardust.

This same process has happened billions and billions of times in our galaxy and is still happening today, with new stars and solar systems still being born. The stars in the sky are all balls of burning energy, with us able to see the resulting light from so very far away.

Not all stars are the same though. Some have noticeably different colours and brightness. This is down to their size, the type of star they are (such as yellow dwarves like our Sun or red giants such as Betelgeuse) and at what stage of their lives they are at. Stars live and die; they don't last forever. On a human scale though, it will feel as if they do. Some have brief violent lives and others last many billions of years. Our own star, the Sun, is just an average star known as a yellow dwarf. If you were on an alien planet looking up it would appear just as many other stars do – a white, ordinary-looking dot.

The Sun is currently halfway through its life and will, in a few billion years, evolve into a massive red giant star, engulfing the planets closest to it, and possibly the Earth. Many stars live and end this way, but some much hotter or larger stars can end their lives in a spectacular explosion known as a supernova, the brightest and most violent event in the universe.

How far?

When we look up at the stars, we see them as if they are painted onto the sky, and ancient cultures believed this was the case. They are actually so far away that we have to measure their distance in 'light years'. One light year is the distance light travels in one year. Light travels at 186,000 miles per second, give or take, so to put that in context, if you were travelling at the speed of light you could travel around the Earth 7.5 times in one second. Now imagine the distance you could cover in a year – in one light year. Impossible distances.

If you travelled in a spaceship at the speed of light (we can't travel that fast yet), then the closest star to Earth would take four years to reach, and the farthest visible star to the naked eye would take around 4,000 years to get to. Therefore, the light from the closest star (not the Sun) has taken around four years to get to us and the light from the farthest star four thousand years. That's quite a 'wow' concept to get your head around, isn't it?

The farthest object visible to the naked eye is actually the Andromeda Galaxy, at 2.5 million light years away – our intergalactic neighbour. On the doorstep, so to speak. You've heard me mention 'galaxy' a couple of times now. Think of a galaxy as a massive swirling island of stars and dust. Some contain hundreds of billions of stars, most with planets orbiting around them, and there are countless galaxies out there. The universe is vast. The galaxy we are in – our home galaxy – is called the Milky Way, and is home to our solar system and our planet, Earth, where we live.

Shapes in the sky

Now that we understand a little more about where and what the stars are, in relation to ourselves, let's start looking at how to find our way around the night sky.

As you scan the heavens you can clearly see that the stars form shapes or groups on the skydome above. Some are clear and obvious, and others are more difficult to see. Remember that they change position throughout the course of a night as the Earth rotates, and are visible or not depending on the time of year and the annual cycle of the Earth's position (the circling of the lightbulb).

These shapes appear to be marked out like a cosmic dot-to-dot, or weird stick men shapes, with some of course standing out more brightly than others. The groupings are known as constellations and asterisms. We often just use the term constellations for everything.

Since ancient times, different cultures have used the stars to make shapes in the sky. Some are recognisable across different civilisations, like the modern constellation of Orion. Some such as those based in ancient Chinese astrology and astronomy are very different to others. This is basically due to the cultural and religious beliefs of the specific civilisation and their time and distance from others. Many constellations and much of the sky lore are derived from the ancient Greeks, who were the first civilisation to develop astronomy to a highly advanced level compared with other cultures. They wove their Hellenistic belief system, as well as the maths and science of the time, into this new and more comprehensive style of stargazing. You will notice a lot of names for constellations stem from Greek and Roman mythology – the dominant civilisations of earlier times – and we will use some of these stories to help us explore and navigate the night sky later in this book.

The entire sky area is divided up into constellations – you may have heard of Ursa Major, Ursa Minor, Orion, Capricorn, Leo, Sagittarius and Andromeda (yes – all of the 'star signs' are in there). The constellations are groupings of stars or an imaginary 'box' of space containing stars within, and tend to have the more serious/astronomy names. In comparison, asterisms aren't actual constellations themselves, but tend to be smaller groupings that make up a noticeable shape within a constellation, or spanning a few different constellations. The most famous asterisms are the Big Dipper and the Little Dipper, the Plough and the Southern Cross.

There are 88 constellations in the night sky, which are unequally split between the northern and southern hemispheres. Not all 88 are visible from both hemispheres though – there are some constellations that can only be seen from either the northern or southern hemispheres.

Look at any star chart or map from a book shop, and you will see that the brightest stars have odd names – like Betelgeuse, for example, which is part of the constellation of Orion. These are usually from ancient Arabic or other early cultures, as a range of peoples would have looked at the same bright stars and built their tales around them.

This is especially the case in the northern hemisphere. The ancient Greeks and Romans didn't venture into the southern hemisphere and their night sky wasn't mapped until European explorers started to chart the southern constellations many hundreds of years later. Then, they named the southern constellations after animals they encountered, or tools they used, such as Tucana – the Toucan, and Sextans – the Sextant (a navigational instrument used by sailors and mariners).

You will notice on some star maps or planetarium apps that constellations and asterisms are overlaid with pictures of the character or object they play host to – animals or figures perhaps. Think of the constellation as the frame or skeleton of dots (stars) joined up with lines and the pictures overlaid. This is known as constellation art and can be very helpful in visualising and remembering the shape of the constellation figure.

Star charts and apps

I recommend you look at star maps and charts as you learn more about the sky above. You will notice that star charts, apps, and astronomy software such as Stellarium show hugely detailed maps of the sky, and many stars with weird symbols and numbers next to them. They include not just stars and planets, but also show the positions of galaxies and nebulae – what astronomers call 'deep sky objects'. The maps have grids, lines, shaded areas and much more.

To someone who has never looked at a star map in detail before, it all looks very complex and confusing. But this is just a chart showing points of interest, landmarks and information, just like any other map. Someone who lives in London and uses the Tube regularly will find the London Tube Map instinctively easy to follow. A country bumpkin like me who rarely visits London may find it complex and confusing. It's got nothing to do with our intelligence, just our level of familiarity, and is something we can improve with practice, so don't be afraid of star maps and spend some time getting used to their layout and display.

Time to look up

Putting the star maps aside, let's try some simple stargazing.

One of the most striking and probably well-known northern hemisphere asterisms is the Plough or Big Dipper, also known as the Saucepan. The names can change in different countries. In the UK, it tends to be the Plough, the Big Dipper in the US, and is also referred to as the Saucepan because of its shape. It's actually an asterism, and makes up the most noticeable part of the large constellation of Ursa Major – also known as the Great Bear – in the northern hemisphere. All you have to do to find it is look around in a northerly direction – it's visible all year round. It is so clear and easy to spot due to its huge saucepan shape. For the beginner in the northern hemisphere, this is the best place to start your exploration of the night sky.

In the southern hemisphere the most noticeable constellation is Crux – the Southern Cross, which looks like a small crucifix formation of bright stars. It's a great starting point for navigating the southern sky, boasting the third brightest star in the night sky and closest star system to our own. At just under 4.5 light years away, the star system of Alpha Centauri is close by, in the constellation of Centaurus. Are there stargazers on a world there looking back at us?

The Plough and Southern Cross are very useful indeed in finding your way around the night sky as they are visible all night and all year long. They are part of a circular group of constellations known as the circumpolar constellations because, yes, they surround the North and South Poles. There are nine circumpolar constellations in the northern hemisphere: Ursa Major, Ursa Minor, Cassiopeia, Cepheus, Perseus, Lynx, Draco, Camelopardalis and Auriga. In the southern hemisphere the circumpolar constellations are Centaurus, Crux and Carina.

You don't need to memorise these constellations but they are helpful as they never set and are always visible at night. Their orientation changes as the night and year progresses but you can always find the North Star at their centre of rotation.

Sometimes I see new sky watchers finding the night sky overwhelming them. Don't let it. Take small steps and go slowly. Remember – this is not a race to memorise and locate the most constellations.

The Plough

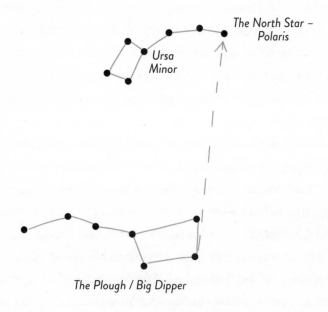

The Plough / Big Dipper

Let's start our journey in the stars with this prominent asterism. The Plough is part of the Ursa Major constellation, and you can use the Plough/Big Dipper to easily find the North Star – Polaris.

As you look at the Plough in the sky (or on the chart below), it will appear as a simple 'pan' shape with the handle extending three stars away from the bowl to the left. Now look at the two stars that form the right edge of the pan – these are known as the Pointers. Join them together with an imaginary line and then extend this line upwards from the pan for a considerable distance, approximately four times the length, until you get to an ordinary-looking brightish star. This is the North Star – Polaris – and will be hugely important in helping you get your bearings while stargazing. (Contrary to popular belief, the North Star is not the brightest star in the night sky. That title goes to Sirius, which we will come to later.)

Congratulations – you've located your first 'constellation' and the North Star. Now, whenever you venture out on a clear night, whether for a night-time ramble or stargazing expedition, the Plough or the Big Dipper is something you'll be able to easily spot without a chart or a telescope. That feels good, doesn't it? This simple connection to the night sky with just the naked eye makes most people feel warm inside (and also rather clever and smug!).

For extra brownie points, let's test your eyesight. If you look at the 'handle' of the Plough/Big Dipper, you will notice two stars very close together if you have good eyesight. The centre star of the handle is called Mizar and another star, Alcor, seems to be right next to it. It's a famous 'double star', Alcor being the faintest. It has been said that in ancient Roman times, these two stars were used as an eye test for Roman soldiers. If you could see both, you could be an archer, with all the privileges that came with the role. Could you have been an archer in the Roman army? Take the test yourself and see.

The southern hemisphere

If you're in the southern hemisphere, finding south isn't as easy as in the north. Look for a small bright constellation of four stars in the shape of a crucifix. Found it? - this is the small but key constellation of Crux – the Southern Cross. Now, draw an imaginary line down its length and toward the horizon to locate the rough position of south.

Exploring further

For part of this and the following chapters, you may benefit from having access to a star map, chart or app to use alongside the charts here in the book. You can continue comfortably without one, but a night sky map, chart or app may help you follow along and visualise the layout more easily. Eventually you'll not need one, but they are always useful for reference and to help jog your memory from time to time.

'Star hopping'

Having identified our first asterism, you can now use the Plough/Big Dipper and the North Star to navigate your way around much of the night sky using a process called 'star hopping'. For example, if you extend the imaginary line you just made from the Plough to the North Star a little further, roughly half the distance again, you come to a star that sits at the tip of what appears to be the gable end shape of a house – a square with a pointy roof on top. This is the asterism of Cepheus.

Next to Cepheus you will notice a striking group of stars that form a 'W' shape – which is Cassiopeia. In Greek mythology, Queen Cassiopeia and King Cepheus were the vain parents of Princess Andromeda. And so you find the constellation of Andromeda by using her parents – the constellations of Cassiopeia and Cepheus.

Draw an imaginary line from the star at the tip of the house shape of Cepheus through Cassiopeia and on again for the same distance to the middle of four stars that appear to make an imaginary curved line. This is the constellation of Andromeda. It's not very obvious and becomes easier to spot with practice. If you have incredibly dark skies, or you could try this with binoculars, you may have spotted a faint fuzzy patch just before the curved line of stars. This is the Andromeda Galaxy – the farthest object visible with the naked eye, which we'll come back to later.

Another way to find Andromeda is by following the same method as we did to find Cepheus from the Plough and Polaris, and then continuing for the same distance again to the bright star at the end of the line of four stars of Andromeda, which is called Alpheratz, and it is also a corner of what appears to be a large square shape. This is the asterism known as the Square of Pegasus (which is part of the constellation of Pegasus – The Winged Horse).

This is at the 'head' end of Andromeda, but if we go to the other end, or her feet, following down the curved length of four stars and next to Cassiopeia, we find a large curved 'Y'-shaped collection of stars. This the hero Perseus brandishing the severed head of Medusa ready to save Princess Andromeda from Cetus, the Sea Monster.

You can also use the Plough/Big Dipper to find Perseus by using the bottom left and top right stars of the pan and drawing a line through them continuing on roughly another four times, and this will take you straight to Perseus.

There are many different routes to take via star hopping. Some are well-known and some are personal to the stargazer. You will find your own preferences.

The power of storytelling

As I mentioned above, most of the northern hemisphere constellations get their names from ancient Greek or Roman mythology and other early cultures. If you are familiar with some of these myths, you can help tell the stories by using the sky. I do this on the top of a windswept hill on my Night Sky Tours and it makes them so much more magical. The myth I just referred to was the famous story of Andromeda and Perseus, and I recommend you explore and enjoy them as you learn to find your way around the sky – *Mythos* by Stephen Fry is a favourite read of mine, as is the *Penguin Book of Classical Myths* and *The Greek Myths* by Robert Graves.

You'll find many constellations mentioned in well-known stories are close to each other and grouped in the night sky, as with the story of Andromeda and Perseus. Legends say that the main characters were placed in the heavens as a reminder or as monuments of their deeds by the gods. Conveniently, these stories or myths help you commit the constellations to memory, helping you to remember their positions in the night sky. The myths also make the act of stargazing a little more fun and interesting.

There are a number of famous constellation groupings or families for both the northern and southern hemispheres. Some, like the myths of Perseus, Hercules and Orion, are grouped together with other constellations depicting characters or objects from myths. Others, like the zodiac constellations, which I'll come back to later, form other groups of constellations that are grouped together for more practical purposes, such as finding the Sun, Moon and planets. These groupings 'organise' the constellations and, if you know the stories or are at least familiar with them, can also help you navigate around the sky more easily.

In this chapter we've begun to familiarise ourselves with the night sky, and I hope now that the stars, the constellations and the shapes they form seem intriguing and no longer daunting. You've now learned the Plough/Big Dipper – the best place to start navigating the night sky – and how to find the North Star from there. Using the basic principle of star hopping, we jumped over to other constellations steeped in mythology, which can also help us navigate to constellations and asterisms representing related characters or objects nearby.

Now, not all of the night sky or objects within are visible at any one time during the year so let's find out more about that next before we launch into our explorations of the night sky, season by season.

Chapter Four

THE SEASONS

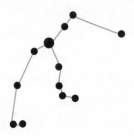

The seasons affect the appearance of the night sky just as much as they do our surroundings here on Earth. What we see above us changes with the seasons, as does our weather and temperature. Humans have always celebrated (and still do) as we herald in each season in various ways. Now that more of the world has central heating and air conditioning plus ready supplies of food, it's easy to underestimate how significant the changing of the seasons would have been to our ancestors – when surviving the winter was an achievement or a good harvest season could mean the difference between life and death.

The seasons also have a major effect on our wellbeing, changing our state of mind and the way we feel. Everyone has their favourite time of year – whether they prefer hot weather, winter snuggling or the transitional seasons of spring and autumn. During the dark winter months particularly, some miss the long hours of sunshine and suffer from Seasonal Affective Disorder (SAD), which can bring on depression and anxiety (largely due to the shorter hours of sunlight). This could possibly be why everyone goes crazy for the beach on the first hot day of spring. It's not just about the day length and temperature though; we're also reacting to the cycle of the natural world around us. The warmer months thrive and burst with life and in the colder months nature appears to slow down and sleep, creating an infinite cycle and ambience that can be very enriching while you stargaze.

An ever-changing cycle of sights, sounds and smells to help you relax, gather your thoughts, unwind and re-charge while exploring the heavens above and natural world around you is certainly part of the stargazing experience. I dislike the cold, but I still enjoy winter and the winter sky. Some say it's the best time of year for astronomy or to stargaze due to the extended hours of darkness. The skies are darker in the winter, but I relish every season as there is always something different to see and experience above us. If you like to stargaze in the summer, then go ahead; if you prefer the long nights of winter, that's great. It's all about doing what you want to do, when you want to do it.

Seasonal stars

As the seasons change, you will notice what you are watching in the sky move, as well as sunrise and sunset times changing. Sometimes we have to wait for constellations or objects to come into view – night sky objects will slowly move out of view as others appear. I have a favourite constellation in each season – they are like old friends I haven't seen for a year. Observing and experiencing them again always makes me happy.

In Chapter 2 we explained the motion of the night sky and why some stars and constellations are seen annually within a particular season; for example, the constellation of Leo is only seen in the spring, the constellation of Cygnus appears just in the summer, Andromeda and Pegasus in the autumn and Orion in the winter. The southern hemisphere has its own constellations and objects visible each season. Over the following four chapters, we discover the highlights of the night sky in each season in more detail.

If we look back at how the seasons are created and what this means for stargazing, there are two details to remember: the tilt of the Earth's axis plus its orbit around the Sun. The Earth is tilted slightly on its axis and so, as the Earth moves around the Sun throughout the year, the poles are either pointing slightly towards or away from the Sun (see diagram below). The easiest way to imagine this is by using a pen and a ball of clay or Play-Doh as mentioned earlier.

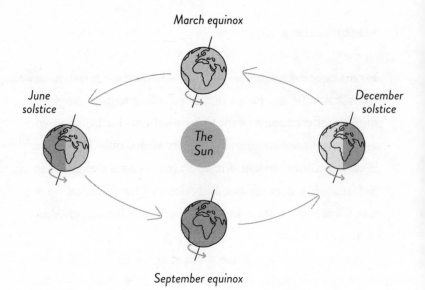

In summer, the North Pole points toward the Sun in the northern hemisphere, and at the same time in the opposite hemisphere the South Pole is pointing away from the Sun, making it winter. Mid-summer and mid-winter are twelve months apart from each other in both hemispheres. These midpoints are known as 'solstices', and at the time of the solstices the Sun is at its highest in the sky during summer and at its lowest point during winter. (The higher or lower energy that reaches the planet due to the angle of the Sun causes the difference in temperature.) It's the same for spring and autumn, they are opposite to each other depending on whether you are in the northern or southern hemisphere.

During the spring and autumn, there is an 'equinox' in late March and September, which is the midway point between spring and summer, or autumn and winter. At the equinox, day and night are of equal length as the Sun is over Earth's equator ('equinox' coming from the Latin for equal and night). As the Earth moves around the Sun, we see a slightly different part of the sky each night. Each season we see the night sky move or change by about one-quarter (or 90 degrees) of the skydome above us, as stars disappear to the west and 'new' constellations arrive in the east. As we move around the Sun in one complete orbit (a year) the cycle starts again.

• • •

With this in mind, let's take a tour of the seasons and the key constellations and highlights of what we see in the sky at that time, so you can navigate and enjoy the night sky above you right now.

As I emphasised earlier, the key to doing anything outside is to be comfortable. Make sure you are warm so that you can stay outside long enough to enjoy the benefits of stargazing. Slow down, relax and take it all in. If you follow this approach, you will successfully stargaze all year round, and truly experience the wonders and sensations of the world around you with each season.

Chapter Five

THE SPRING NIGHT SKY

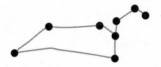

The seasons are crucial for life on Earth and have a fundamental effect on the natural world around us and the sky above. This is certainly true in the spring as we leave the cold and dreary sleepiness of winter behind and life begins to literally 'spring' out of the ground with temperatures slowly rising and the weather becoming more pleasant.

Birds and animals prepare to raise new families as animals that have hidden away over the winter months, such as insects, become more active. We humans feel the positivity of spring as we spend more time outside, getting the garden ready, visiting parks and enjoying outdoor pursuits.

Of course, as with any season, each hemisphere is opposite to the other so for this chapter we will be looking at the spring night sky in the northern hemisphere and the autumn night sky in the southern hemisphere – choose the section appropriate for your position. We'll also see some objects that are common to observers on both hemispheres at the same time.

Use the accompanying star charts to help you to spot and move around the constellations – especially at the beginning as you learn to recognise the main stars and shapes.

Spring equinox

Officially (and astronomically), spring begins on the spring equinox (also known as the vernal equinox), which occurs in the northern hemisphere around 20 March, and it ends with the summer solstice on 21 June. At the same time, the autumn equinox and winter solstice occur in the southern hemisphere.

As you experience the sights, sounds, smells and atmosphere of spring around you in the daytime, look for the same at night-time as the creatures of the night prepare for and go about their nocturnal activities. It will have started to get darker later in the evening and the temperatures aren't quite as harsh as in the winter so it's a great time to start stargazing, but the night-time conditions in the spring can be very chilly with frosty nights earlier in the season, so have your warm clothing and sturdy footwear to hand.

Northern Hemisphere – Spring

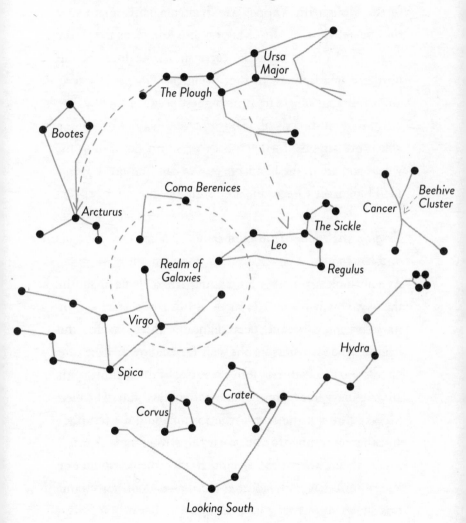

If you remember, we have already looked at the asterism of the Plough/Big Dipper in Chapter 3. This appears in the constellation of Ursa Major – also known as the Great Bear – and can be seen all year round from most of the northern hemisphere. In the spring the Plough is placed almost directly above us at its highest point in the sky. We call the point directly above us the 'zenith', and this point will move slightly further north or south depending on where you are in the hemisphere. We can find north using the Plough/Big Dipper.

Finding the Plough/Big Dipper and North

As you look up at this large saucepan or ladle shape in the sky, you will see it is made up of seven stars – four stars creating a bowl shape and another three forming the handle. The two stars of the pan furthest away from the handle are the Pointers. Join these two stars together with an imaginary line from the base of the bowl (the star called Merak) through the star at the top of the bowl (Dubhe) and then continue this imaginary line roughly four times its length and you reach a brightish star – the North Star or Polaris. You have now located the direction of north, and this allows you to get your bearings.

It's always a good idea to find the North Star first. From here you can quite easily find east, west and south directions. To find south, look behind you when facing the North Star, or turn to the left or right 180 degrees if you want to be a little more scientific about it. It's very useful if you are stargazing in unfamiliar surroundings and need a quick bearings check.

Leo (The Lion)

Next, we can use the Plough to find one of spring's main constellations: Leo, the lion. It is named for its sphinx-like shape, and after the Nemean lion killed by the hero Heracles in the first of the twelve labours he was forced to undertake according to Greek myth (Hercules in Roman mythology). The Nemean lion was a formidable beast and its likeness was therefore placed in the heavens by the gods and we see it as the constellation of Leo today. In the story, the lion was a ferocious monster with razor sharp claws that were harder than steel and a pelt impervious to mortal weapons. It terrorised the local people of Nemea and was eventually cornered and killed in a cave by Heracles, who then cut off its pelt using its own claws and wore it for protection as a cloak and to make him appear even more mighty.

There are two simple ways of finding Leo. The first is to face south about mid-evening and look for its sphinx-like shape: a backwards question mark of stars known as the 'sickle' forms the lion's head and mane. Its body stretches out to the left of the sickle, appearing as if it is crouching or in the loaf position, with its legs folded and tucked beneath (like a cat).

The second way to locate Leo is to find the Plough and

imagine the bowl of the ladle is full of water. At the base of the bowl drill an imaginary hole and imagine the water pouring out and falling for roughly five times the depth of the bowl until it reaches and splashes onto the lion's back.

Leo is one of the larger and more striking constellations, and its prominent shape can usually be found just by scanning the sky. We'll look at more of the zodiac constellations in Chapter 10, but for now this is a great start to a constellation that boasts some striking objects. There is a double star at the base of the question mark (called Regulus) that is divisible by large binoculars or a fairly good telescope, and some distant galaxies, including the Leo Triplet, beneath the base of Leo's body. Use a planisphere, printed night sky map or a night sky app (with a dark adaption feature) to help you find these fainter objects with your binoculars or a telescope.

If your stargazing interest leads you into hunting galaxies, the spring night sky is certainly for you. The area between Ursa Major (the Plough or Big Dipper area), Leo and the faint constellation of Coma Berenices and Virgo to the left (or east) of Leo is your playground. There are literally dozens and dozens of galaxies here to be viewed with small telescopes. This is called the Virgo Cluster of galaxies, of which our own Milky Way is part (see Chapter 9).

If we use the constellation of Leo as a central point, we can further explore the spring night sky quite easily.

Cancer (The Crab)

To the right of Leo, you will see the faint constellation of Cancer (the Crab), also one of the 12 constellations of the zodiac. This constellation looks like a crab on its side in constellation drawings but this is a challenge for most to see as it comprises only five stars, which aren't all that bright. Look for its upside down 'Y' – a star map may help you find it more easily. Cancer is home to one of the highlights of the spring sky – the Beehive Cluster (or Praesepe, meaning 'manger' in Latin). The Beehive is a beautiful open cluster of stars, visible as a faint smudge to the naked eye near to the centre of the constellation, and one of the closest star clusters to Earth.

In Greek mythology, the constellation of Cancer relates to the second of Heracles' twelve labours. A Hydra (sea monster) was sent to kill Heracles (Hercules) by Hera, queen of the gods. As Heracles seemed close to defeating the Hydra, Hera did her best to sabotage the battle in favour of the multi-headed monster by distracting Heracles with a giant crab. (Hera really hated Heracles and wanted rid of him as he was the result of her own husband's adulterous behaviour with a mere mortal woman.) Heracles paused for a moment in battling the Hydra, turning his attention to the crab and crushing it underfoot before finishing off the Hydra. Hera was enraged that Heracles was again victorious and saddened by the death of the crab and the Hydra. Due to their bravery, she placed both the crab and the Hydra in the heavens close to Leo so they could continue to tell the story for all eternity.

Hydra (The Sea Monster)

The constellation of Hydra is actually the largest constellation in the night sky. It depicts a gigantic one-headed water snakelike monster. In Greek mythology the Hydra had many heads and if you chopped a head off, two more would instantly grow back in its place. Constellation art shows it only with one head, but you can imagine more. Hydra literally snakes across the sky just below Cancer, heading south-west below Leo and then the faint southern constellations of Sextans – the Sextant – and the Crater (the Cup).

Hydra's snake-like tail then rises up slightly under the constellation of Corvus (the Crow) and the large constellation of Virgo. Depending on your location, part of the body of Hydra may disappear entirely beneath the horizon.

Corvus (The Crow)

The constellation of Corvus is small and unremarkable but is easily identified by its asterism in the shape of the sail of a boat: a skewed rectangular shape of four stars. Corvus was a white raven sent by its master, the Olympian god Apollo, to watch over one of his mortal lovers. Eventually, the lover fell for a mortal man and Apollo was enraged at Corvus for not intervening, turning its white feathers black with scorching anger. It is said that this is why all ravens are now black.

Virgo (The Virgin)

We now move upwards away from Hydra and Corvus towards the bright star Spica and its home in the constellation of Virgo – the Virgin. Virgo is the second-largest constellation in the night sky and, of course, another zodiac constellation; it represents Dike or Dice, the goddess of justice in Greek mythology. If you look at its constellation art, this often depicts an angelic winged female holding an ear of grain in her left hand. This is the position of the very bright star Spica – Latin for 'ear of grain'. As you look at the constellation, it appears to be on its side with the head end to the right of Spica and legs stretching to the left. Spica lies about one-third of the way along the constellation with the head end stars forming a large bowl. This area is, as I mentioned above, a favourite place for galaxy hunters and is home to the Virgo Cluster of galaxies or Realm of Galaxies.

Boötes (The Herdsman)

As we star hop around the spring sky, we look up from Virgo and see an incredibly bright, orange-coloured star. This is Arcturus – the brightest star in the northern hemisphere, a massive red giant star 36 light years away at the foot of the large kite-shaped constellation of Boötes (pronounced bow-OH-tease) – the Herdsman.

Boötes is unremarkable apart from Arcturus at its tip. Arcturus is often referred to as the star that heralds spring and can be found from the familiar asterism of the Plough/

Big Dipper. As you look at that asterism, look to the left of its bowl and follow its handle. See how the stars arc? Use this arc and continue it for twice the distance again of the handle until you reach Arcturus (guardian of the Bear in Ancient Greece).

Arcturus, Spica (in Virgo) and Regulus (the brightest star in Leo), form a large asterism of three bright stars, known as the Spring Triangle.

In myth, Boötes represents Arcas, son of Zeus and Callisto, a follower and close companion of the goddess Artemis. Zeus fell in love with Callisto and transformed himself into the likeness of Artemis and took advantage of her. As a result, she bore a son – Arcas. Zeus really liked the ladies and it got him and the subjects of his desires, and their resulting offspring, in all sorts of trouble with his vengeful and spiteful wife and family. Fooling around with the king of the gods did not grant you a 'get out of jail free' card. The other Olympians would do their best to cause the utmost and everlasting misery to those who Zeus adulterously spawned or bedded.

Artemis was infuriated and felt betrayed at her closest companion breaking her vow of chastity and so banished Callisto while adopting Arcas, her own half-brother. Hera, Zeus's wife, was also angered by her husband's ongoing inability to keep it in his trousers so to speak and followed Callisto into the forest and turned her into a terrifying she-bear, in which form she remained as her son grew up. Arcas was told that his mother had been killed by this ferocious bear and he swore to one day track down and kill the bear to avenge his mother. As he grew, Artemis trained him in the skills of archery and hunting and he became a very handsome and formidable warrior. He was out hunting in the forest one day on his own and was spotted by the bear-shaped Callisto. Overcome with emotion and love for her long-missed son, she ran towards him to finally embrace him. However, all Arcas saw was a furious bear charging toward him. He raised his bow and loosed an arrow to slay the charging animal – his mother!

Zeus was watching his love and his son while staying in the shadows and saw Callisto charge and Arcas loose his arrow. With a mighty cry and bolt of lightning he deflected the arrow and sent both Callisto and Arcas into the stars where they could be together forever. Today we see them both as Ursa Major – the Great Bear (Callisto) and Boötes (Arcas).

Boötes in other stories can be seen driving his great plough across the stars and thus maintaining the rotation of the sky.

This same myth can also include Ursa Minor – the Little Bear or Little Dipper and home to Polaris – in some versions, swapping Boötes for the Little Bear. There are many versions of the stories of the skies, I like the Little Bear version myself.

Cephus and Cassiopeia (The King and Queen)

Let's now turn our attention to the north. Find the North Star again using the Plough/Big Dipper and you can then locate the constellations of Cepheus and Cassiopeia by extending the imaginary line for a similar distance again. Cepheus looks like a house shape and is quite faint. Cassiopeia is the slightly smaller but brighter 'W'-shaped constellation to its left.

These are King Cepheus and Queen Cassiopeia, the boastful parents of Princess Andromeda. Their vanity and arrogance brought the fury of the gods down upon them with a sea monster ravaging their lands, and the only way to pay penance was to sacrifice their daughter Andromeda to the sea monster. However, the hero Perseus turned the sea monster into stone with the severed head of Medusa, thus saving the day for Andromeda and her parents. Unfortunately, this didn't work out as Hera, queen of the gods, was outraged at the death of (another) pet and the arrogance of the king and queen. She placed them both in the sky on their thrones where, once a day, they would receive a daily dunking as their heads dip into the sea (as seen from Ancient Greece) as a final and perpetual punishment and humiliation.

Indeed, if you look at Cepheus and Cassiopeia in the spring night sky and consult their constellation art, you will see both are upside down and close to the horizon. If you are looking at them from Greece or the Mediterranean area, they will definitely get their heads wet.

Southern Hemisphere – Autumn

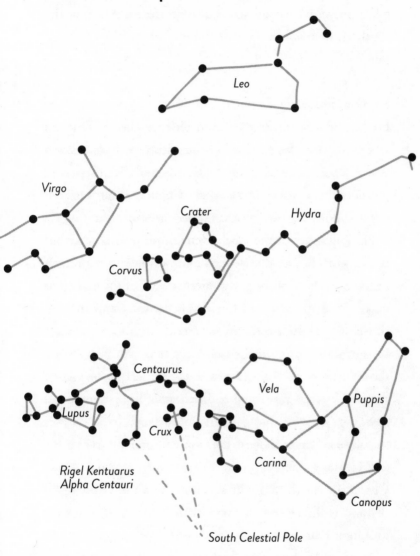

Leo

Virgo

Crater

Hydra

Corvus

Centaurus

Vela

Puppis

Lupus

Crux

Carina

Rigel Kentuarus
Alpha Centauri

Canopus

South Celestial Pole

Looking South

You will recall that when it is spring in the northern hemisphere, it's the opposite and therefore autumn in the southern hemisphere, a great time to observe a wealth of highlights.

Finding South

First let's get bearings. And to do that we need to find out where south is. In the northern hemisphere, the northern celestial pole is visible. But in the southern hemisphere it is below the horizon so we need to look for the southern celestial pole. Everything has flipped around, so to speak.

Finding south in the southern hemisphere isn't as easy as in the north but it's doable if you know how. First, scan the heavens for a small bright constellation of four stars that make the shape of a crucifix. This is the small but striking constellation of Crux – the Southern Cross – the smallest constellation in the night sky. Next, look for two stars a short distance to the left of Crux. The left/lower star of the two will be very bright. This star is Rigil Kentaurus, meaning 'Foot of the Centaur' in ancient Greek, as it is literally the right foot in the constellation of Centaurus (the Centaur). The star is more commonly known as Alpha Centauri – the third brightest star in the whole of the night sky and is the closest star system to our own, at just over four light years away.

Now, go back to the Southern Cross (Crux) and draw an imaginary line down its length and toward the horizon. Then go to the centrepoint between Rigil Kentaurus and the slightly fainter star above and draw an imaginary line toward the horizon from there also. The point at which the two imaginary lines meet is the celestial south pole – the direction of south. After doing this, you should be able to get your bearings in the southern hemisphere.

Upside down

Many of the constellations mentioned above when we were looking at the northern hemisphere are visible in the northern sky of the southern hemisphere at the same time. The northern circumpolar constellations won't be visible though and will be below the northern horizon to some extent, depending on your location. The visible constellations we covered above will, however, appear upside down if you are viewing from somewhere like Australia, or South Africa. The constellation of Leo will have Cancer to the left and Virgo above and to the right with Boötes below and are easily identifiable. Above Cancer, Leo and Virgo are the smaller constellations of Sextans (the Sextant), Crater (the Cup) and Corvus (the Crow). Directly overhead near Corvus is the tail of Hydra – the largest constellation in the night sky. In the southern hemisphere, it appears with its body stretching down to its head in the lower parts of the north-west.

Centaurus (The Centaur)

Looking south around mid-evening, we see the Southern Cross (Crux) high in the sky. The large constellation of

Centaurus (the Centaur) envelops Crux to the left and from above. The constellation represents a centaur – a mythical half-man, half-horse creature, with the torso and head of a man and the body of a horse. In Greek mythology, Centaurus represents the wise centaur Chiron, the son of a Titan and a sea nymph. He lived in a cave and taught many of the well-known heroes, such as Jason, Achilles, Heracles and Theseus, among others. Chiron was accidentally struck by one of Heracles' poisoned arrows (dipped in the blood of the Hydra) during a skirmish between Heracles and a group of centaurs in a disagreement about some wine, believe it or not. He was immortal and could not die, but suffered tremendous pain from the poison. To ease his suffering Zeus placed Chiron in the sky and, to the naked eye, Centaurus is a bright constellation displaying a recognisable shape of the centaur.

As I mentioned earlier, Centaurus is home to Alpha Centauri or Rigil Kentaurus – the closest star system to ours – at the front leg/foot of the constellation. The lower half of the constellation crosses the Milky Way and is full of clusters of stars, including Omega Centauri – the brightest and largest in the Milky Way. There are also some great nebulae and galaxies to hunt for here.

Lupus (The Wolf)

Centaurus is often shown holding and attacking an animal to his left. This is the constellation of Lupus (the Wolf) representing a wolf ready to be sacrificed on the altar of the constellation of Ara, found to the bottom left of Centaurus and Lupus.

The Argo (The Ship)

To the right of Centaurus, we find the constellations of Puppis (the bow or poop deck of the ship), Vela (the sails) and Carina (the hull or keel) of the ship. These three constellations make up the shape of the old constellation of Argo Navis or the *Argo*, the mighty ship that Jason and the Argonauts sailed in on their quest for the Golden Fleece. Indeed, constellation art shows the ship combining all three constellations and it is one of my favourites in the southern night sky. The constellation of Carina includes Canopus, the second brightest star in the night sky (outshone only by Sirius) and also contains the Eta Carinae nebula at the stern of the ship.

• • •

The southern hemisphere sky has so many wonders to explore, they deserve a book in their own right. Whether you're looking at the main constellations of spring in the northern hemisphere or autumn in the southern, locating these constellations frequently will enable you to identify them from memory quickly. Don't try to learn everything at once, just take your time and enjoy it. The more you skygaze, the more familiar the arrangements of the stars will become and easier to lock into your memory.

As spring starts to transition into summer in the north

and autumn to winter in the south, the night sky changes also, as we will soon see.

Chapter Six

THE SUMMER
NIGHT SKY

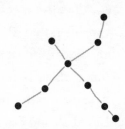

You'll either love or hate the summer night sky; it can be like Marmite for some. It's the season with the longest days and shortest nights, which can mean less time stargazing for some, or a whole night out under the stars for others. For those who like to venture outdoors and enjoy all things alfresco, it's a great time of year so, putting gourmet tastes aside, let's do some summer stargazing.

Again, with any season the hemispheres are opposite each other, so we will be looking at the summer night sky in the northern hemisphere and the winter night sky in the southern hemisphere in this chapter. Summer in the northern hemisphere begins on the solstice around 21 June and ends with the autumn equinox. At the same time, it is the winter solstice in the southern hemisphere. Some objects can be seen by observers in both hemispheres simultaneously, so we won't be doubling up or repeating ourselves too much.

Summer for me and many others is a favourite time of year. It's warm, it's light and it makes you feel happy, with long days of sunlight and the world around you filled with life. The summer night sky is a marvel to look at and enjoy, with so much to see. It is the season where the night sky truly comes to life and is the most accessible season to see it (if you have pleasant weather, that is).

The one negative is that it doesn't get suitably dark until much later in the evening, which is annoying for astronomers who like to go to bed early. In mid-summer, it may not get suitably dark at all and can never reach much darker than twilight in some more northern parts. The further north you go, the less darkness or night there will be. Some locations near and above the Arctic Circle remain in perpetual daylight for many weeks as the North Pole is then tilted towards the Sun and constantly illuminated by its light.

Further south (even if the sky isn't truly dark), there are still wonderful sights to see with the naked eye. Many amateur and professional astronomers prefer the winter's prolonged darkness, but you just have to time it right to enjoy all that the summer has to offer.

Northern Hemisphere – Summer

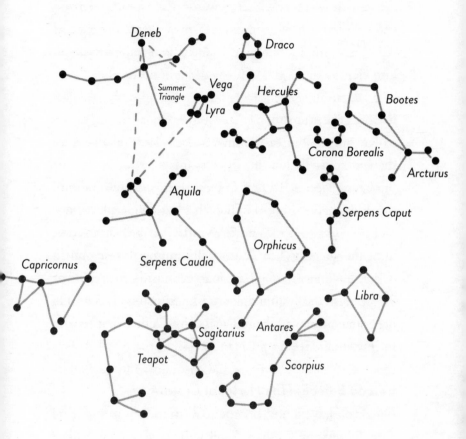

As we transition from spring into summer, the spring constellations move closer towards the western horizon and become less evident, and are then eventually lost. At the same time, the summer constellations and objects rise into view.

Boötes (The Herdsman)

In late spring and early summer, look for the large kite-shaped constellation of Boötes (as mentioned in the previous chapter). To find Boötes, start your journey with the Plough/Big Dipper. Find the large saucepan shape and its curved handle and if you draw an imaginary curve using the stars in its handle, you will continue to a very bright orange-looking star. You have found Arcturus, the main star in Boötes, at the bottom tip of its large kite shape of stars. For more about the myth behind this constellation, turn back to the previous chapter.

Corona Borealis (The Northern Crown)

Once you have found Boötes, look to the left of it to find a small bowl- or 'C'-shaped collection of seven stars; this is the constellation of Corona Borealis (the Northern Crown). As with most of the constellations, there is a Greek myth associated with it.

The Northern Crown represents the crown worn by Ariadne. Ariadne helped the Greek hero Theseus escape from the Minotaur, the half-man half-bull monster who lived in the Labyrinth (which is said to be beneath the island of Crete). After saving Theseus by giving him a ball of string to find his way out of the Labyrinth, Ariadne was then abandoned on an island afterwards by him. (What a complete git, as we say in England.) There, she was spotted by the god Dionysus, who married her. The crown to which this constellation relates was made by the god Hephaestus and worn by Ariadne on her wedding day. It's one of the more pleasant endings in Greek mythology and a big fat raspberry in the direction of Theseus. It's a small constellation but rather charming given its happy ending, and it sits close to one of the main summer constellations . . .

Hercules

The constellation of Hercules (or Heracles in Greek mythology) can be found to the left of Corona Borealis. It's not an exceptionally bright constellation considering the fame of the hero it depicts, but can be found by looking for its wedge-shaped 'keystone' asterism that forms the torso shape of the hero. Another way of finding Hercules is by looking for the bright star Arcturus (at the base of Boötes) as above, and the bright star Vega. Hercules lies in between these.

He was son of the king of the gods, Zeus, and a mortal woman, Alcmene, and the great-grandson of the hero, Perseus. Heracles was mighty and famous, being the pinnacle of manliness and bravery (which was a big deal in Ancient Greece) while an icon in Greek mythology –

the archetypal flawed hero made good and made immortal in the end. He is most famous for undertaking his twelve labours: twelve impossible tasks that he was forced to undertake as a blood debt after (accidentally) killing his family in a fit of madness plotted by the goddess Hera. In Hellenic times, blood debts could be paid off by the pardon of a god or king and required the offender to do particular and usually impossible or long-lasting tasks as penance.

The night sky references many of these labours and the story of Heracles features heavily within many of the northern constellations. If nature charities existed back then, they would have had a field day with Heracles given the number of endangered or unique animals he despatched. He even stole the king of the underworld's three-headed dog!

The constellation is noted for being home to the Great Globular Cluster – M13, the brightest and most well-known cluster of stars in the night sky.

You may have noticed I also called it 'M13'. The 'M' stands for 'Messier' or, more accurately, French astronomer Charles Messier who in the 18th century catalogued 110 deep sky objects such as galaxies, nebulae and clusters into the Messier Catalogue. You will see objects on most star maps with an M and a number from 1 to 110, and most of these are visible with binoculars or a small telescope. If you have ideal conditions and eyesight, you can see M13 in Hercules with the naked eye as a smudge of light within the keystone shape at the centre of Hercules.

Hercules' constellation art shows Hercules kneeling and facing the constellation of Draco, the subject of his eleventh of his twelve labours . . .

Draco (The Dragon)

One of these doomed magical beasts lies directly above Hercules at his crouched feet and is the constellation of Draco. Draco (Latin for dragon) is a large and not-very-obvious circumpolar constellation – its body snakes from its head then towards Cepheus and back again, and between Ursa Minor (the Little Bear) and Ursa Major (the Great Bear). Draco represents the dragon Ladon who was killed as he protected Hera's golden apple tree when Heracles stole one of the golden apples as one of his twelve feats. Hera placed Ladon in the sky afterwards as the constellation Draco.

In the summer, the head of Draco is directly above the zenith (right above your head). The bright stars Eltanin and Rastaban represent his eyes while two other stars form the lozenge-shaped asterism of Draco's head. Another way of finding the lozenge of Draco's head is to draw an imaginary line from Aquila's bright star Altair through to Vega in Lyra and onto the lozenge and Draco's bright eyes. We will learn about Altair and Vega shortly.

Ophiuchus (The Serpent Bearer)

Look below Hercules to find the very large constellation of Ophiuchus (the Serpent Bearer). The constellation represents Apollo's mythical son Asclepius, who was famed for his healing skill and power, and depicts a man holding a snake coiled around his waist, made up of two separate constellations – Serpens Cauda (the Snake's Tail) and Serpens Caput (the Snake's Head) – with Ophiuchus between them.

Ophiuchus is one of the less obvious constellations, but its constellation art will help you learn to identify its shape. (The constellation is famed for being the unofficial thirteenth constellation of the zodiac.) Asclepius did many great deeds and his healing powers rivalled those of the gods. Zeus killed Asclepius as Zeus and his brother Hades – the god of the underworld – feared that Asclepius would make men immortal thus making them equal with the gods. He was placed in the stars as recognition of his good deeds and skill.

Libra (The Scales)

Beneath Ophiuchus, we find three more constellations of the zodiac.

The constellation of Libra (the Scales) can be found near the horizon to the left of Virgo (see Chapter 5) and represents the scales Virgo (Dike/Dice) uses to administer justice. It is a southern hemisphere constellation but is visible from northern hemisphere mid-latitudes such as the southern UK and below. It takes the shape of a triangle with the base representing the weighing beam and two lower stars as the weighing pans – constellation art will help you identify it more easily.

Scorpius (The Scorpion)

Immediately to the left of Libra is the very striking southern constellation of Scorpius (the Scorpion) with its curving scorpion tail shaped like a fishhook. Scorpius is easily found as it lies near the centre of the Milky Way and has a distinctive scorpion shape. Its most distinctive feature is the star Antares – the Heart of The Scorpion – a massive red supergiant star. Close by, and to the right of Antares, three stars form the right claw and help to identify this low constellation. It's so low that only its top half can be seen from mid to high locations such as the UK.

Scorpius represents the scorpion sent to kill Orion by Gaia (Earth) for boasting he could kill every beast on Earth while he was out hunting with the goddess Artemis. Scorpius and Orion are now opposite each other in the sky; as the Scorpion rises in the east, Orion can be seen fleeing as he sets over in the west.

Sagittarius (The Archer)

To the left of the tail of Scorpius, we can hop over to the southern constellation of Sagittarius (the Archer).

Sagittarius lies in the Galactic Centre (the rotational centre) of the Milky Way and is due south mid-evening in summer. It is a striking constellation representing a centaur – a half-man half-horse mythical creature pointing a bow and arrow towards the heart of the Scorpion. For the full effect, look at the constellation art for this complex figure. To help you locate it though, look at it on a star chart first as Sagittarius has an unmissable asterism at the bow and arrow end that immediately identifies it as a giant teapot.

This area of the sky is a haven for deep sky object hunters with a plethora of nebulae, clusters, galaxies and stars to be found if you want to dust off your telescope. There is so much to see here you can run out of time before the sky starts to lighten or those elusive objects dip beneath the horizon.

Capricornus (The Sea-Goat)

To the left of sagittarius is the final zodiac and very faint constellation of Capricornus, (the Sea-Goat), best seen late summer. In Greek myth, Capricornus represents Pan, the god of the wild with the legs and horns of a goat – one of my favourite mythological creatures. Pan was placed in the sky as an act of thanks by the gods for his help during the war of the Titans and the attack of Typhon when he blew his conch shell to scare away the Titans. Typhon was a massive and powerful hundred-headed fire breathing serpent and the most feared monster in the world, hell-bent on ruling the cosmos. (Even the gods feared him.) Pan also alerted the gods to the monster and helped them disguise themselves as animals so that Typhon would ignore them and they could escape. Pan himself jumped into a river and turned his lower half into a fish to swim away to safety. Constellation art from a map or star atlas will help you identify the constellation in the shape of a rough triangle.

Aquila (The Eagle)

Now, look away and higher in the sky to the constellation of Aquila (the Eagle), above Sagittarius and to the right of Ophiuchus. Aquila represents the eagle that carried the thunderbolts of Zeus. In other stories, Aquila snatched up the Trojan boy Ganymede and took him to be an immortal cupbearer for Zeus on Mount Olympus.

Again, it's a striking constellation, resembling a bird with wings outstretched and the bright star Altair at the head of the constellation flying towards the constellation of Cygnus (the Swan), which is coming the other way, higher in the sky. The two celestial birds appear to be flying towards one another.

Cygnus (The Swan)

Cygnus's constellation above Aquila is one of the most prominent and easiest constellations to identify in the night sky and also one of the most beautiful. It looks like a swan in flight, with a formation of bright stars in a large cross that make up the swan shape, with the double star Albireo at the end of its long neck and its head and Deneb, the brightest star in the constellation, at its tail. It is otherwise known as the Northern Cross asterism.

Cygnus straddles the Milky Way as it flies toward Aquila. Filled with nebulae and bright stars, it is a favourite astro-photography area, with objects such as the Veil Nebula and the North America Nebula to see, to name a few.

There are numerous myths associated with Cygnus,

one being of Zeus transforming himself into a swan and seducing Leda, the Queen of Sparta, on a riverbank. Instead of giving birth, Leda laid two eggs, the first containing Castor and Helen (who went on to be Helen of Troy), and the second bearing Clytemnestra (who would become wife of Agamemnon), and Pollux (Polydeuces). Castor and Pollux went on to be great heroes and Argonauts from Jason and the Golden Fleece's story.

The second myth refers to the hero Orpheus, the sad Master of Music, placed in the sky as a swan after his death next to his favourite instrument, the lyre – the constellation of Lyra, which was gifted to him by the god Apollo.

Lyra (The Lyre)

Lyra is a small but very striking constellation to the right of Cygnus representing a lyre, a small stringed instrument similar to a harp. The constellation itself resembles a small, skewed rectangle with a very bright star at its upper end. This is Vega, the second brightest star in the northern hemisphere and the brightest star of the summer night sky. Lyra contains a couple of deep sky objects, one being the famous Ring Nebula – M57 – which is worth hunting for in a good telescope.

The Summer Triangle

If you draw an imaginary line linking the constellations of Lyra, Cygnus and Aquila by their brightest stars, Vega, Deneb and Altair, this makes one of the most prominent and obvious asterisms in the night sky – the Summer Triangle. The Summer Triangle straddles the Milky Way and is an excellent guide for finding objects in and around it – a welcome signpost in summer.

Southern Hemisphere – Winter

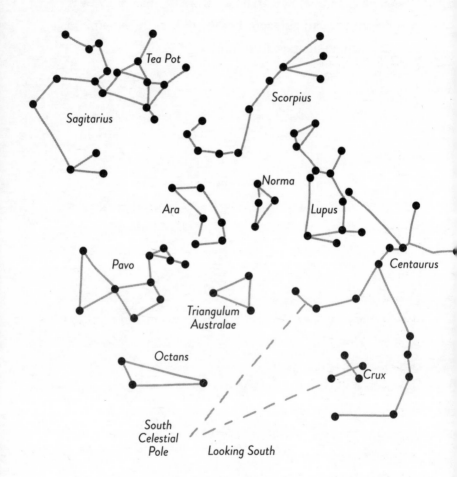

Wintertime in the southern hemisphere begins on the winter solstice in June and ends on the vernal equinox in September.

If you have jumped straight to the southern hemisphere in this chapter, I recommend you read the text above as many of the northern sky's objects are visible to both hemispheres. I'll flag these as we work through this section.

Hercules and Draco

Start by finding your bearings, as we covered earlier in this book. To locate north, look low on the horizon mid-evening for the constellation of Hercules, which is the right way up as seen from the southern hemisphere; this is at north.

To find Hercules, use the bright star Arcturus (of the kite shape of Boötes) to the left of Hercules and quite low near the horizon. Draw an imaginary line to the right until you come to another bright whitish-blue star – this will be Vega (in Lyra, and the brightest star of the summer sky). Hercules can be found midway along this imaginary line.

Cygnus, Lyra and Aquila

Rising from the north-east horizon to Hercules' right are the constellations of Cygnus, Lyra and Aquila, forming the Northern Summer Triangle (called the Winter Triangle in the south, of course) with the Milky Way passing through it. You will notice that the Milky Way is much fainter at this point but gets brighter the higher in the sky you go and further south you are. The Milky Way is at its best in the southern hemisphere where much of it can be seen in all its splendour.

Ophiuchus (The Serpent Bearer)
Above Hercules, and to the left of the Summer Triangle and Milky Way, lies the large constellation of Ophiuchus and the two Serpents. From this part of the world, though, he appears to be upside down and standing on his head.

Scorpius and Sagittarius
As we look directly overhead at the zenith, we have the Scorpius and Sagittarius constellations with the centre of the Milky Way passing through them. Southern hemisphere observers are blessed with this view as it is partially hidden in the thick low atmosphere in the north. Both Scorpius and Sagittarius can be seen in full with the fishhook asterism in Scorpius on full display along with all the deep sky wonders in this area.

Corona Australis (The Southern Crown)
Corona Australis (the Southern Crown) is nestled beneath the 'Tea Pot' asterism in Sagittarius.

If you now turn around and face south, high in the southern sky Centaurus (the Centaur) still dominates along with the tiny but distinctive constellation of Crux (the Southern Cross).

Going further

To the left of these, we have a collection of small constellations that will benefit from being studied and viewed using a good star map. They consist of Norma (the Set Square or Carpenter's Square), Ara (the Altar), Pavo (the Peacock) and Triangulum Australe (the Southern Triangle). There are more constellations to see as we venture closer to the south celestial pole but they are small so I recommend you spend extra time tracking them down with a map to explore further.

• • •

And with that, we have covered the summer sky in the north and winter in the south. It is a great starting point to look for smaller constellations and deep sky objects if you wish to take your stargazing further.

As the season moves on, the constellations move more westerly each evening and eventually give way to autumn's constellations in the north and those of spring in the south. The weather begins to change, as do our outdoor and seasonal activities. Some look forward to a change of season, some want the summer to last forever. The one thing that can be guaranteed is that the night sky's appearance will change dramatically with new sights coming into view, as we will explore in the next chapter.

Chapter Seven

THE AUTUMN NIGHT SKY

Leaving the long days of summer behind, autumn brings noticeably shorter hours of daylight to feed plant life and warm our surroundings. The natural world is preparing for the long sleep of winter. One day fruits and grains are ripe for picking and you are in shorts and a T-shirt, and the next, you need to dress for weather considerably cooler. It's not just the difference in temperature; the weather is changing and all around plant life is dying off or shutting down, ready for winter.

It is time to migrate to warmer climes for some creatures after spending the summer feeding, growing and preparing. In late summer and early autumn, we mark harvest time for the human and natural world. As the season moves on, much of the fruit has been eaten, picked or fallen, and the leaves on trees and bushes are a flash of red, yellow and orange as they begin to fall.

For some, the change can be pretty depressing. Others find the vibrant colours of nature quite breath-taking. It's a time to take stock, and to look towards Halloween, Thanksgiving and Christmas. I live in rural Oxfordshire in England and look forward to stacking logs for the fire and going through all the traditions and activities autumn has to offer. I believe it to be a very positive time of year. The seasons aren't just about the change of weather and the natural world; they are about your wellbeing also. If you embrace the seasons and what they bring to your life, you will find the change easier to bear.

The night sky is always an amazing sight, though, whether you are with friends, family or on your own. Whatever way you choose to stargaze, you will notice the autumn sky getting completely dark much earlier in the evening.

In this chapter, we will be looking at the autumn night sky in the northern hemisphere and the spring night sky in the southern hemisphere. Autumn in the northern hemisphere begins on the autumn equinox in late September and ends on the winter solstice in December. At the same time, spring begins and ends in the southern hemisphere.

Both hemispheres have constellations that can be seen from the other at the same time so I recommend you read the entire chapter, wherever you live.

Northern Hemisphere – Autumn

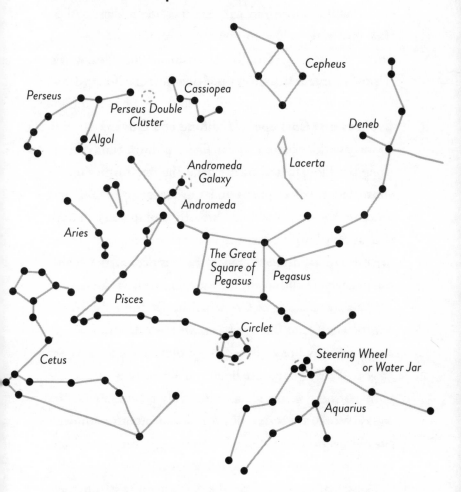

As we scan the night sky, the Summer Triangle and summer constellations move further west and finally disappear as they give way to the constellations of autumn. Looking north, we see the familiar asterism of the Plough/Big Dipper in its classic upright position near the horizon.

Cepheus and Cassiopeia (The King and Queen)

As we have done before, we can find the North Star (Polaris) using the Plough, and then extend the line further to the constellations of Cepheus and Cassiopeia (the King and Queen). You can also find them by looking right above you, as they both sit at the zenith mid-evening at this time of year. (As mentioned earlier, the position of the zenith can change depending on the time and your location.)

The constellation of Cepheus isn't far from the North Star (Polaris), and the main stars within it form the shape of the gable end of a house. To the right of Cepheus is the bright 'W'-shaped constellation of Cassiopeia. These are both northern circumpolar constellations and sit on the Milky Way, and because of this, they are awash with deep sky objects for those with telescopes to go hunting – this is especially the case with bright Cassiopeia.

Constellation art shows the pair sat upright on their thrones, and this is where we start with the stories of the constellations in the autumn sky. Use these tales to help you find your way around if that appeals.

Cepheus was the king of Ethiopia, and his wife was Cassiopeia. They were the proud parents of Princess

Andromeda, the fairest maiden in the land. Cassiopeia was overheard boasting that her and her daughter's beauty was greater than the Nereids – the sea nymphs. At this, the Nereids complained to the god of the sea – Poseidon – who sent Cetus the sea monster to terrorise and ravage the land as punishment for their boastfulness. The gods and immortals were easily offended and wrathful in their vengeance, it would seem!

King Cepheus consulted an oracle, a kind of seer or prophet in Greek myth, and was told the only way to stop the monster rampaging on his coastline and lands was to sacrifice the princess to the sea monster. Believing that this was the only way to save his kingdom, Cepheus ordered his daughter to be chained to a rock for Cetus to devour.

As all this was about to happen, the hero Perseus was returning from slaying the Gorgon Medusa. Mounted on Pegasus with the head of Medusa in a bag, Perseus noticed the young maiden chained to a rock with a sea monster bearing down on her. Perseus was instantly smitten with Andromeda as he gazed upon her. He swooped down with Pegasus and pulled the head of Medusa out of its bag. Cetus caught the gaze of Medusa, who even in death could still assert her petrifying magic, and the sea monster instantly turned to stone. Andromeda and the lands were saved.

This is a very brief version of the story, but it's enough to help familiarise ourselves with the autumn night sky and help us find our way around.

Andromeda (The Chained Woman)

We've mentioned already that Cepheus and Cassiopeia are directly above, near the zenith, in the autumn. Face south and look overhead or directly above you. Now, draw an imaginary line from the star at the tip of the house shape of Cepheus to the brightest star in the 'W' shape of Cassiopeia, and carry it on to the next bright star, which appears in a long train or string of four stars forming a curve. This is the constellation of Andromeda.

Andromeda is an unremarkable constellation, looking like a curved line of stars, with two more faint stars slightly above the second star from the right forming her waist. The constellation art of Andromeda will show the chain of stars running the length of her body, with her head to the bottom right and feet to the upper left as seen when looking south.

There is one perfect jewel of the autumn sky that we passed on our way from Cassiopeia, and that is the Andromeda Galaxy – M31. This is a stellar island of a trillion stars and the furthest object visible to the naked eye at 2.5 million light years away.

Pegasus (The Winged Horse)

Another way of being sure you are looking at the constellation of Andromeda is to look for the 'Great Square of Pegasus' asterism that uses the brightest star in Andromeda – Alpheratz – at its top left corner. Pegasus is quite a large constellation easily identified by its large square shape. It is upside down as we look at it in the autumn and the constellation depicts the front half of the flying horse. There are offshoots of stars above and to the right for his two front legs and below and to the right for his neck and head. Pegasus is striking and easy to find without the need to star hop.

Perseus (The Hero)

To the left of Andromeda, and below Cassiopeia, is the upside down curved 'Y'-shaped constellation of the hero – Perseus. It is a reasonably large and bright constellation with a long, curved chain of stars representing Perseus's body. The constellation branches off at the constellation's brightest star, Mirfak, continuing down and to the right to a collection of stars representing Medusa's head, with the bright star Algol – the Demon Star – representing one of her eyes.

There are numerous deep sky objects in Perseus due to its proximity to the Milky Way, including one of my favourites, the Perseus Double Cluster. Perseus is also famous for the Perseid Meteor Shower, which occurs annually in August (see meteor showers in Chapter 11).

Aries (The Ram)

We now look below Andromeda and past the small constellation of Triangulum to the three autumn zodiac constellations. Directly below Andromeda and Triangulum and to the right of Perseus lies the constellation of Aries (the Ram). The constellation is tricky to spot, with only one bright star in the line of four stars that make up the length of the body and head of the ram. A star map with constellation art will help you identify it.

In Greek myth, Aries the Ram (not Ares the god) was a flying ram with a golden fleece that performed a daring rescue but was then sacrificed to the gods. The dead animal's golden fleece was then placed in an oak tree. (I'm not sure this was the reward the ram deserved, in my opinion.) Eventually, the magical golden fleece was the goal of Jason and the Argonauts on their epic quest.

Pisces (The Fishes)

To the right of the constellation of Aries, and below Andromeda and Pegasus, is the large constellation of Pisces (the Fishes). A large 'V'-shaped constellation, it represents two fish attached by the tail with a cord or ribbon. The constellation has two distinct spurs representing the fish, and at the end of the righthand spur or westernmost fish is a circular asterism. In Greek mythology, the two fish represent Aphrodite and her son Eros (Venus and Cupid in Roman stories) after they were transformed into fish by water nymphs so they could swim away and escape from Typhon. Luckily Aphrodite and Eros escaped, and Zeus eventually defeated Typhon with his thunderbolts.

Aquarius (The Water Bearer)

To the right of the constellation of Pisces and below Pegasus is the third zodiac constellation on our autumn tour, Aquarius (the Water Bearer). Aquarius represents the boy Ganymede who was swept up by Aquila the Eagle and brought to Olympus and became an immortal cupbearer for the god Zeus. Aquarius is a reasonably large constellation of three widely spaced chains of stars. The first to the right represents an outstretched arm; the central chain forms the body of Aquarius and his right leg. The easternmost chain to the left represents water being poured out of a jar. His outstretched arm joins his body at his head/shoulder near the brightest star in the constellation. Just to the left of this, and at the top of the chain of stars representing pouring water, is the asterism of the 'Water Jar' or 'Steering Wheel' – I believe it looks more like a steering wheel. As with most of the constellations, a star chart with constellation art will help you identify this constellation more easily.

Cetus (The Whale, or Sea Monster)

My final highlight for the autumn sky tour is to the left of Aquarius and below Pisces – the constellation of Cetus (the Whale, or sea monster). We heard about the deeds and demise of Cetus in the myth of Perseus and Andromeda above. The constellation is large and does indeed resemble an animal with a long, crouched body, long neck, and round head to the left of the constellation. This is the asterism known as the Head of the Whale. A star chart with constellation art may help, but with this constellation, I think it may not be necessary.

Southern Hemisphere – Spring

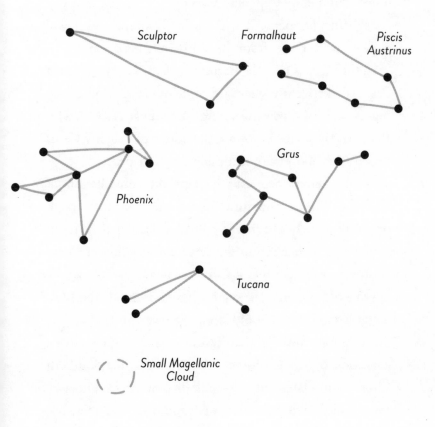

Sculptor

Formalhaut

Piscis
Austrinus

Phoenix

Grus

Tucana

Small Magellanic
Cloud

Facing South

If you face north from the southern hemisphere, most of the constellations mentioned above will be visible – the exceptions being Cepheus, Cassiopeia, Perseus and the Plough/Big Dipper. All the northern constellations will appear upside down from the southern hemisphere

Andromeda is near the horizon, and the Great Square of Pegasus takes centre stage and pride of place. As you look higher in the sky, you will see Aquarius above Pegasus, with Pisces and Aries lower toward the northeast. Mighty Cetus towers high above with its upside-down shape.

At the zenith overhead is the faint constellation of Sculptor. We can now turn around and face south. High overhead is the bright star Fomalhaut (meaning the Mouth of the Fish in Arabic), in the small constellation of Pisces Austrinus – the Southern Fish. Fomalhaut also marks the left foot of Aquarius. The myth associated with this constellation is similar to that of Pisces, as above.

As we look lower in the sky and further south, we see a collection of avian constellations known as the Southern Birds – the first being the constellation of the Phoenix, the mythical bird that catches fire and then rises anew out of its ashes. This constellation is pretty faint, but you can make out the phoenix shape. To the right of Phoenix is the constellation of Grus (the Crane). This isn't a bright constellation, but you can quite clearly make out a wading bird shape. Below both of these, we come to Tucana (the Toucan), another faint constellation. Tucana is the home of the Small Magellanic Cloud, a small satellite galaxy of the Milky Way.

Many of the southern night sky constellations are faint at this time of year but hunting them down can be very rewarding. The same goes for any season and either hemisphere.

• • •

There is much more to see with the naked eye than listed in this chapter, and a detailed star map will help you explore further. We've looked at the prominent constellations and objects of the autumn to help you get started and acquainted with the night sky at this time of year. As always, in both hemispheres, the following season beckons and the next, for some, is the season to be jolly.

Chapter Eight

THE WINTER NIGHT SKY

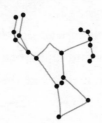

Winter is the coldest and harshest time of the year when much of the natural world sleeps or scratches out a living to survive. The reason it is so cold is simply down to the fact that it is the time of year with the least amount of sunlight. In winter, the Earth's tilt causes the winter pole to point away from the Sun, resulting in less sunlight and warmth reaching us. Because of this, most plant life struggles to survive, with deciduous trees like oak, ash and elm being bare and many plants appearing dead and in a deteriorated state of entropy. The plants have shut down and have gone to sleep, so to speak, to save energy and live off the stores they built up in the warmer months. Some die back or even die off completely, repeating a yearly circle of life. Many birds have migrated to warmer climes, and the dawn and dusk chorus is now more of a lullaby than the cacophony of sound as heard in the warmer seasons, especially during spring. Mammals such as bears, hedgehogs and dormice have gone into hibernation in their warm dens and nests.

Where I live in the UK, we see the traditional Christmas card animals such as deer, pheasants, foxes and owls taking centre stage in the countryside, as we do with the remaining squabbling birds on the garden feeders. It is usually cold, wet or windy, or a combination of all simultaneously, and some in the human world find it pretty miserable. I expect the animal world does also, but they just get on with it as many humans do. However, just like any other season, winter has lots to see, do, and experience, and much of this is good for you: walking on a windswept hill in the cold air, jumping in puddles, or playing in the snow are good for the soul. A roaring log fire, a steaming hot bowl of soup and thick woolly socks are some of the things I love about winter. In the northern hemisphere, the winter solstice is around 21 December. We only get around eight hours of daylight in the UK, starting and ending our working day in darkness. Go further north above the Arctic Circle, and you will be in perpetual darkness for several weeks. In the northern hemisphere, the winter solstice is just before Christmas with all the warmth, joy and tradition it brings. We tend to spend a lot more time indoors in the winter due to the less than pleasant weather but it's the opposite in the southern hemisphere where you can celebrate Christmas in shorts on a warm beach or at a barbecue.

In the northern hemisphere, one outdoor activity that is perfect in winter is stargazing. Due to the fewer hours of daylight, you can stargaze to your heart's content all night long if you are dressed well to stay comfortable for a cold night under the stars. I always recommend a flask with a hot drink or soup, and if you are lucky to have a chimenea or firepit, you can light this and make it a real occasion. I've even heard of people stargazing from the warm waters of their hot tubs with ice and snow around them. I'd love to try that myself one day. Most importantly, enjoy what you are doing and your time stargazing.

Again, please read the whole chapter if you are in the south, as much of the northern hemisphere is visible at the same time as the southern.

Northern Hemisphere – Winter

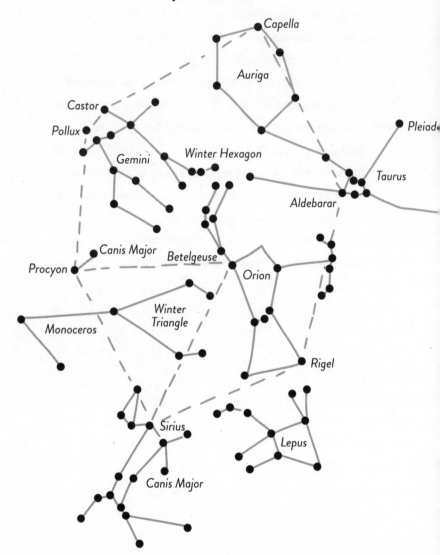

Facing South

The winter night sky is breath-takingly beautiful, and to many stargazers and astronomers, it is their favourite time of the year – for the views and the extended amount of darkness. Due to their striking appearance, all we need to do is face the right direction to notice and identify some of the winter constellations. This is especially true of the mighty constellation of Orion (the Hunter), the most prominent constellation in the winter night sky. Some constellations require the help of others to star hop to, and we will look at these shortly.

Orion (The Hunter)

Face south and roughly halfway between the zenith and the horizon is the large shape of the Orion constellation. The main shape of Orion resembles a tunic pulled in at the middle by three bright stars, or a giant butterfly on its side. The three stars close together form Orion's belt (also known as the Three Kings or Three Sisters). A very bright orange star, Betelgeuse, is to the top left of the tunic, marking the right shoulder. At his left shoulder is the star Bellatrix, and at the bottom of his tunic, as you look at it, is the very bright blue-white star on the right, Rigel. The star to the left is called Saiph.

This makes up the large and pronounced main body of the constellation, but there is more. To the right of Bellatrix, Orion is stretching out his left arm and holding the pelt of a dead animal. This appears as a faint chain of six stars in a curve. You could mistake this for a bow as I did when I first started stargazing. At Orion's right shoulder – left as you are looking at him – is Betelgeuse, above which is his extended right arm and club made from six faint stars rising in an arc, completing the overall shape of Orion in the sky.

If we go back to the three stars that make up his belt – Alnitak, Alnilam and Mintaka – you will see a chain of stars that make up his sword just below. If you look a little longer, you may notice a faint smudge. In binoculars or a small telescope, this becomes more than a smudge; it becomes the Great Orion Nebula (M42) – the closest and brightest star-forming nebula to us, truly massive with hundreds of stars being born within it.

Now look carefully at the stars of Orion's sword, his belt and a star an equal distance to the right, and they form a sizeable mirror-like shape known as the Mirror of Venus.

Betelgeuse is probably one of the most famous stars in the night sky and a prime candidate for what is known as a supernova – when a massive star at the end of its life collapses under the weight of its own gravity, resulting in an enormous explosion. Supernovas are the most violent events in the galaxy and can be seen for weeks at a time. It is believed that Betelgeuse will look like a second tiny sun in the daytime and cast shadows in the evening when it does explode. Some believe this may have already happened, and the light from the exploding Betelgeuse just hasn't reached us yet. All we can do is keep a watch on it and wait and see.

In Greek mythology, Orion was the son of the god

Poseidon and mortal Euryale, the daughter of King Minos of the island of Crete. He was extremely handsome and grew to be a famous and mighty hunter. He fell in love with the Pleiades, the seven sisters and daughters of the Titan Atlas and Oceanid Pleione. One myth says the Pleiades were eventually placed in the sky just out of reach of Orion to stop him from getting to them. In another tale, Orion is hunting with the goddess Artemis and her mother, the Titan Leto, and boasted that he could kill all the animals in the world. Gaia – the Earth goddess – overheard him and sent a scorpion that killed Orion so he wouldn't carry out his boast. Both Orion and Scorpius – the Scorpion – are placed in opposite parts of the sky as Scorpius chases Orion, as we noted in Chapter 6.

Sirius (The Dog Star)

We can use the constellation of Orion to navigate our way around the night sky quite easily as it partly forms the Winter Triangle and Winter Hexagon asterisms. Before we come to those, however, we must introduce another of the night sky stars.

Look to the left below Orion by using the three stars in his belt and drawing an imaginary line through them, and going down and to the left, you will reach an incredibly bright star. This is Sirius (the Dog Star), the brightest star in the whole night sky. It's also known as the 'twinkling star' or 'multi-coloured star' as it shimmers and displays so many different colours, due mainly to its position as seen from the northern hemisphere. It is shallow in the sky, so we see it through the thickest part of the atmosphere with all the turbulence and pollution concentrated therein, distorting the light of the star. Sirius's brightness can amplify this, but all stars twinkle depending on their brightness and position. Stars high overhead at the zenith twinkle the least compared to those closer to the horizon as the atmosphere is thinner directly above.

Canis Major (The Big Dog)

Sirius is known as the Dog Star as it is found in the constellation of Canis Major (the Big Dog), a reasonably bright and conspicuous constellation showing the shape of the dog.

Lepus (The Hare)

To the right of Canis Major, and beneath Orion, lies the small constellation of Lepus (the Hare). Not particularly striking and rather faint, this constellation benefits from having constellation art to help identify it from a map or chart. Lepus appears to be running away from Orion and his hunting dogs.

Monoceros (The Unicorn)

Above and to the left of Canis Major is the very faint constellation of Monoceros (the Unicorn), another that is not particularly easy to spot, but a star chart will help.

Canis Minor (The Little Dog)

Carry on through Monoceros until you get to the bright star Procyon in the constellation of Canis Minor (the Little Dog), parallel to Orion's shoulders. You can find Procyon by drawing an imaginary line through Orion's shoulders and continuing it left until you reach the bright star. If it weren't for the bright star Procyon, the constellation would be very hard to spot.

Canis Minor is often thought of as one of Orion's hunting dogs. In another story, it resembles an uncatchable fox pursued by Canis Major – the fastest dog in the world. The fox, however, could not be caught, so Zeus, king of the gods, turned them both to stone and then placed them in the heavens.

The Winter Triangle

Orion, Canis Major and Canis Minor, with their stars Betelgeuse, Sirius and Procyon, make up the Winter Triangle, appearing as a prominent equilateral triangle-shaped asterism.

The Winter Hexagon

We can now make out the shape of the huge Winter Hexagon asterism as we hop around the constellations of the northern winter sky.

Gemini (The Twins)

Draw an imaginary line from Sirius to Procyon and then

carry on for a similar distance but to the right slightly until you come to the first of two bright stars higher in the sky. The first star you come to is Pollux, and the second somewhat higher star is Castor, and they represent the heads of the twins in the zodiacal constellation of Gemini.

Gemini is a fairly large constellation with the bright stars Castor and Pollux being the most obvious feature. Imagine Castor and Pollux, arms around shoulders, posing for a photo as their bodies extend toward Orion. In myth, they were the sons of Zeus (in the shape of a swan) and Leda, Queen of Sparta, as mentioned in Chapter 6, and were key figures in the story of Jason and the Argonauts.

Auriga (The Charioteer)

Next, draw an imaginary line from the star Pollux to Castor and then draw another to the right for quite some distance to a bright star almost overhead at the zenith. This is Capella, the brightest star in the slightly elongated pentagon- or hexagon-shaped constellation of Auriga (the Charioteer).

Capella is a beautiful star and the sixth brightest in the night sky. As well as the winter months, Capella can be seen shimmering low on the northern horizon in late spring and early summer, accompanied by eerie noctilucent (night-shining) clouds. There are different myths associated with Auriga and most involve a charioteer either driven by Hephaestus, the smith god and creator of the chariot, or Erichthonius, the first man to train horses and lead a chariot. Zeus and the other gods were so impressed with this feat, they placed him in the heavens. Constellation art shows him holding the reins of a chariot with one arm and a goat and kids with the other.

As we continue around and form the Winter Hexagon asterism, we draw an imaginary line from Capella past the small triangular group of stars representing the goat kids and the pale orange bright star in the bottom right of the Auriga constellation. If you are using a star map or app, the name of this star is Hassaleh. Travel through Hassaleh for a similar distance again to a very bright orange star in a 'V'-shaped collection of stars. This is Aldebaran, in the constellation of Taurus (the Bull).

Taurus (The Bull) and the Pleiades Cluster

Taurus is a reasonably large and bright zodiacal constellation with a long thin 'Y'-shaped collection of stars forming the bull's front legs, head and long horns. The bright 'V'-shaped group of stars at the centre of this 'Y' shape is known as the Hyades Cluster and represents the bull's head, with Aldebaran as his eye. Only the head and forelegs are depicted in the constellation, which can be better understood in constellation art.

To the bottom right of Taurus, we see the familiar shape of Orion with three bright stars in his belt. Draw an imaginary line through all three stars and continue to Aldebaran as another way to find the bull.

If we continue this line on some more through Taurus,

we arrive at one of the gems of the winter night sky: the Pleiades Cluster (known as the Seven Sisters or Subaru to Far Eastern astronomers). The Pleiades and Hyades clusters are the closest open clusters to Earth.

Remember that the Pleiades were the seven daughters of the Titan Atlas and Oceanid Pleione, as were the Hyades. For his part in the war between the Titans and Olympians, Atlas was forced to hold up the sky for all eternity and was therefore unable to protect his daughters from the amorous advances of Orion, who pursued them relentlessly.

Zeus felt remorse for his old friend Atlas and swept up the Pleiades and placed them in the sky where Orion would never be able to get to them. If you look at the winter sky, you can see the endless chase of Orion and the Pleiades with Zeus as the bull facing off with Orion and protecting them. Taurus is an old constellation; it looks like the head of a bull and has been found in highly accurate cave paintings. In Greek mythology, Taurus represents Zeus transformed into a bull in true Zeus style, as he loved to transform into animals when pursuing the ladies.

The Winter Hexagon

To complete the Winter Triangle, we draw an imaginary line from Aldebaran to the very bright star Rigel, the brightest star in Orion, and we are back where we started.

This little tour is helpful in taking you round the constellations of winter and forms the Winter Hexagon asterism. The Winter Hexagon is impressive in its own right as you look up at it. If you get the whole shape in your view, it's a beautiful sight. Start anywhere along its path, and you'll find it a handy guide to hopping around the winter constellations.

As you will learn, the northern constellations in winter are bright and probably the easiest to get to grips with and remember. Using imagery and stories helps you to remember, and explore the sky in this season.

Southern Hemisphere – Summer

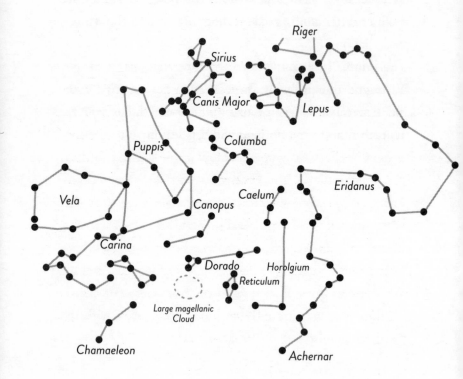

Facing South

The warmth of the southern hemisphere is a wonderful contrast to the wintry skies of the north – but shares many of the constellations and features.

The Winter Hexagon

Facing north and low near the horizon lies the very bright star Capella, as we mentioned in the northern hemisphere. Its proximity to the horizon will depend on your location. From Capella, we can go either left or right and up to chase the route of the Winter Hexagon as above, or the Summer Hexagon as we now call it, as seen from the south. Are you confused yet? Don't be. All we have done is change its name, and quite rightly so as it's summer in the southern hemisphere. Remember that everything is upside down and back to front as seen from the southern hemisphere. To find your way around and learn more, recap with the northern hemisphere if you haven't already.

Sirius (The Dog Star)

Looking directly above at the zenith of the sky, is the brightest star in the night sky – Sirius – with the constellation of Canis Major (the Big Dog) slightly off to the northeast, with Lepus (the Hare) due north along with Eridanus (the River).

Eridanus (The River)

This is a huge constellation snaking its way from Orion westward toward Cetus (the Whale), back towards the zenith and then down towards the south-west, ending at its bright star, Achernar.

Eridanus represents the story of Phaeton, the son of the sun god Helios. Phaeton wanted to drive his father's Sun chariot across the sky, and after much pleading Helios agreed, as long as Phaeton kept to the track. Phaeton got in the chariot and took the reins, and the chariot flew into the sky. Due to his inexperience and the strong horses, Phaeton lost control. The chariot flew erratically and too low to the ground, burning the lands and drying up the rivers and seas, resulting in the Sahara Desert. The whole world was at risk of burning up, so Zeus threw a thunderbolt at Phaeton to end the devastation. Phaeton, on fire, plummeted into the River Eridanus (often identified as the River Nile in Egypt). The shape of the constellation is also said to show the path he travelled before Zeus stopped him.

Going further

As we move further toward the south, we see the smaller constellations of Columba (the Dove), Pictor (the Painter) and Dorado (the Dolphin or Swordfish). To the east, we see the looming hulk of the Argo Navis, representing the ship named after its builder Argus and the goddess Athena, and sailed by Jason and the Argonauts.

• • •

As winter progresses, the constellations of spring start to make their presence known, and the yearly cycle begins again. Each season has its wonders to see and experience, and I thoroughly recommend you choose your favourites in each part of the year. Relax into your skygazing as you gain in confidence, and also enjoy a sense of achievement. You can take as much or as little as you like from each season and the night skies within them. These past four chapters are intended as a starting block to help you begin so that you can learn at your own pace and enjoy stargazing at leisure. If you wish to learn more, that is entirely up to you.

Chapter Nine

THE MILKY WAY

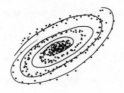

You should feel more confident about what you can find in the skies after the last few chapters. Practice will help: stargazing isn't something you can rush or pick up all at once. The sky is big, and I thoroughly recommend you explore it the way that pleases you the most, at your own pace.

On any clear night, we look up and see stars and sometimes the Moon and planets (which we will cover in the next chapter). We see stars of different brightness and the shapes they make spread across the sky. Over the last few chapters, we looked at the seasons, how to identify some of the groups of stars that form constellations, and finding your way around the night sky in each season. But now, we delve a little deeper – a lot deeper, in fact – into deep space, and we are going to look at something truly beautiful.

Our home galaxy

When you have good clear dark skies, you may notice something other than stars and constellations overhead. Can you see a very faint band of what looks like cloud stretching across the sky? It's not a cloud; it's something else. After your eyes adjust, you will see that it's full of stars. Millions and billions of stars – over 100 billion is an estimate. This is the Milky Way, our home galaxy.

Every star you see in the night sky is within and part of the Milky Way galaxy, including our Sun. So, what is a galaxy? Put simply, a galaxy is an island of millions, billions or even trillions of stars in deep intergalactic space. It is held together by gravity, with a supermassive black hole, millions or billions of times the Sun's mass, at every galaxy's centre, powering it.

Black holes are scary things, not only due to an inescapable fate if you get too close to them, but they are challenging to get your head around. Let me try to explain one to you. A black hole is a region in space usually caused by a collapsing star or other massive body with runaway gravity. This results in the force of gravity becoming infinitely strong. So strong that not even light can escape its gravitational pull, and time even slows down the closer you get the black hole's 'event horizon', the boundary or point of no return where you would forever vanish as you enter the black hole. This is where our general understanding of physics breaks down as we can't see past this point, hence the name black hole.

Black holes are weird, and there is a great deal of speculation about what happens inside them. Do you get crushed to an infinite point as you enter the black hole and added to its mass, or are you spat out into another universe intact? Some theories believe black holes could power wormholes, where you could instantly travel to one part of a galaxy or even the universe in an instant. Regardless, black holes make most people's heads hurt.

The night sky is full of galaxies – hundreds can be seen through telescopes. Some galaxies are giant elliptical blobs resulting from many collisions between themselves, and other galaxies being absorbed into each other. Some are magnificent spiral galaxies, such as many you can see in Hubble images you will find in a copy of *National Geographic*.

The Milky Way is a spiral galaxy, and we know this by the way it appears to us as a band in the sky. We are part of it and either looking away from it or through its centre/disc. The disc of the Milky Way appears as a faint band of cloud or light stretching across the sky from one horizon to the other.

Galaxies aren't just full of stars; they also contain gas, dust and all the ingredients needed to make everything within them: stars, planets, rocks, the elements, life and people (the last part true in the case of the Milky Way). It's improbable there are humans or humanoid life forms in other galaxies, or on planets within our own like there are in movies and TV, where conveniently we all look pretty similar, apart from the odd extra eye or tentacle.

It is now understood that most stars within the Milky Way, and in other galaxies, have planets around them. Astronomers have identified thousands of them in all parts of the night sky, and these are just in our cosmic neighbourhood. It is also believed, and some say a mathematical certainty, that many of these planets can support life. Not just many; there could be millions of planets capable of supporting life in the Milky Way alone.

We are not alone

So, what would life on these planets look like? It depends on many different factors. Some life could be early in its development or have never made it past a puddle of organic goo. Planets could have their versions of dinosaurs or advanced civilisations. At the time of writing this book, we just don't know and don't yet have the technology to look deeper and find out. However, we will soon be launching more significant and better space telescopes that could very well identify planets with life on them. I think it highly unlikely we've been visited by little green men due to the sheer distances involved (and the lack of evidence). I could be wrong though, or just lucky enough not to have had an encounter with one.

The closest galaxy to Earth apart from our own is the

Andromeda Galaxy, in the constellation of Andromeda, as discussed in Chapter 5 and Chapter 7. The Andromeda Galaxy is the furthest object visible to the naked eye at 2.5 million light years away, and visible as a faint oval smudge about the diameter of the Full Moon. You will need a very dark moonless sky to see it.

It is massive, containing around a trillion stars and on a collision course with our own Milky Way as the two galaxies are gravitationally attracted to each other. Andromeda and the Milky Way will collide in an epic display of star birth and light in around four billion years. The two spiral galaxies will repeatedly tear through each other over a few million years, eventually merging into what is known as a giant elliptical galaxy – a massive blob of stars. But let's not worry about that!

Viewing the Milky Way

So how can we see the Milky Way, our home galaxy? Contrary to popular belief, you don't have to go to exotic locations such as the desert or Australian outback, although these will give fantastic views. Some believe the southern hemisphere's view is better as you can see the galaxy's centre in its entirety plus the Milky Way's satellite galaxies, the Large and Small Magellanic Clouds. However, the Milky Way is still quite a spectacle to behold in the northern hemisphere, especially in summer.

The Milky Way is visible from anywhere on the planet but you will need a dark, light-pollution-free sky to see it. Unfortunately, it is nearly impossible or a real challenge to see in urban areas as the light pollution drowns out all or most of its faint light. The light of the Milky Way comes from the millions of stars, some of which are thousands of light years away, and some of which are obscured by large clouds of cosmic dust, creating spectacular nebulae on the edge of human eyesight. Many of these wonders come to life in specialist astronomy images.

If you want to see the Milky Way for yourself, get to an area with little or no light pollution and let your eyes adapt to the darkness for several minutes. You don't need any equipment, just your eyes. Look up and try to spot a band of light that can almost be mistaken for a very light cloud stretching from horizon to horizon.

Below is a guide to locating the Milky Way at the different times of the year; the directions are approximate depending on location and time of viewing. Reading the previous chapters will help you identify many of the constellations if you haven't already. As with the constellations, the Milky Way does not move; the Earth is moving and changing its angle as it passes through the seasons. The position of the Milky Way and the constellations it passes through will change from night to night depending on where you are, the time of night and the time of year.

A star map, chart or planetarium app will also help you just as long as you protect your dark-adapted eyesight. You can also just look up for the Milky Way yourself without any aid or guide. It's entirely up to you. The Milky Way is a wonder to behold with the naked eye, and contains countless objects worth hunting for in binoculars or a telescope.

Northern Hemisphere – Spring

In the spring in the northern hemisphere, the Milky Way is seen stretching from the south-west mid-evening – from the bright star Sirius in Canis Major, up and to the left of Orion in Monoceros. It then passes between Gemini and Orion, arcing north-west through Auriga, Perseus and Cassiopeia, as it dives down into the northern horizon.

Northern Hemisphere – Summer

Rising from the south, the Milky Way is at its brightest at this time of year. The centre of the galaxy is close to the horizon in the constellation of Sagittarius and rises upwards toward the bright star cloud in the constellation of Scutum. It then passes into the constellations of Aquila and Cygnus, as does a band of dust along its plane known as the Great Rift. There are a number of regions and bright or dark spots along the way, and a good star atlas will help you identify them. Once past Cygnus, the Milky Way continues to Cepheus and Cassiopeia before disappearing in Perseus on the northern horizon.

Northern Hemisphere – Autumn

As the Earth tilts away from the Sun in the northern hemisphere, the centre of the Milky Way is hidden, and its faint band appears on the western horizon in the constellation of Aquila. It then rises in an easterly direction, passing overhead through Cygnus, Cepheus, Cassiopeia, Perseus, Auriga and Gemini as it dives below the eastern horizon.

Northern Hemisphere – Winter

In winter, the Milky Way is rising from the south or southwestern horizon depending on when you are looking. It moves up through Canis Major, Monoceros and past Orion to the constellation of Auriga. At this point, the Milky Way is overhead, and we are looking out and away from the galaxy and its faintest stars. From Auriga, we head north through Perseus, Cassiopeia, Cepheus and the small lizard constellation of Lacerta before the Milky Way disappears beneath the horizon.

Southern Hemisphere – Autumn

Rising from the south-east, the brightness of the galaxy's centre can be noted as it rises toward the Southern Cross and the bright star Rigel Kentaurus, otherwise known as Alpha Centauri. It then passes through the extensive collection of constellations making up Argo Navis, clearly visible overhead from a dark location with its Carina Nebula. It then heads north-west through Canis Major and Sirius – the brightest star in the night sky – before reaching the horizon in the north-west.

Southern Hemisphere – Winter

Rising prominently from the south-west, the Milky Way is seen at its best with the entire shape of the galactic bulge and its centre on view. Leaving the sails of the constellation of Vela, it climbs overhead through the Southern Cross, Centaurus and Scorpius with the centre of the Milky Way overhead in Sagittarius. A good star atlas will help you identify many of its dark and light features. It then heads north-east through the bright Scutum Star Cloud onto Aquila before diving below the horizon.

Southern Hemisphere – Spring

During spring in the southern hemisphere, the Milky Way can be seen leaving the horizon in the south or south-west, depending on the time of evening. It rises from Centaurus and heads west through Scorpio, Aquila and Cygnus before disappearing in the north. It doesn't get high in the sky compared to summer and winter, where much of it can be seen overhead

Southern Hemisphere – Summer

During the summer months, the Milky Way rises from the southern horizon, passing through Crux, Argo Navis and then into Canis Major. The Large and Small Magellanic Clouds can be seen prominently in the south. From Canis Major, it passes into the northern constellations of Monoceros, Gemini and Auriga as we look out and away from the galaxy. The Milky Way then descends on and below the northern horizon.

• • •

The Milky Way is a glorious sight but the night sky contains much more than constellations and deep sky objects such as galaxies, nebulae and the Milky Way. It has rocks that move with the seasons and interlopers that can appear over a few days, weeks, months or even a split second. We will look at these visitors to our night sky in the following chapters.

Chapter Ten

THE MOON AND PLANETS

We have covered a tremendous amount about the night sky already, and there has been an awful lot to take in. Stargazing is a little bit like riding a bike. You fall off a lot when you are learning, and then all of a sudden, you find your balance. It takes time, but you get there if you keep practising. The sense of self-accomplishment and satisfaction is more than worth it.

There are no prerequisites or tests at the end of this book, nor should there be any expectations. It's your sky, your leisure time and your enjoyment, so do it the way that pleases you the most. As with learning anything in life, you won't be able to retain everything, and you will forget things, so refer back to this book whenever you are unsure. It's here to help.

Things that move in the night

As we scan the familiar stars and constellations of the night sky when we look up, we can sometimes see things that aren't there all the time. The Moon is the most obvious of these, but there are others too. You'll see stars that seem to change position from month to month or week to week, or aren't usually there. What are these celestial wanderers?

Well, they aren't stars. They are planets.

Planets

The word 'planet' comes from the Greek word 'Planetes', meaning 'wandering star', and each was associated with a different god. Modern science discovered that these weren't stars at all but other worlds in our solar system and, luckily for us, you can see most of them just with the naked eye from the surface of our planet, Earth. (Five planets are visible to the naked eye and another two require a telescope or powerful binoculars to see them.)

Just as Earth does, all planets orbit the Sun, but the solar system is a very diverse place, and each world has its own orbit and its own particular characteristics. We will look at each planet shortly, but we need to understand where to find them first. The planets don't just appear at random in any part of the sky as many people believe. All of the planets, including Earth, orbit the Sun on the same orbital plane.

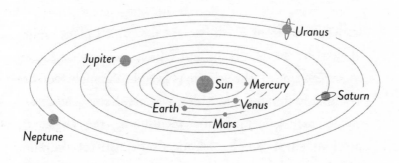

Imagine a flat disc – like a record/LP or similar – and at its centre is the Sun. Each music track on that record could be considered the orbit of a planet, and as you look at the record, especially if it's an album, it has multiple tracks/songs moving out from its centre. Now imagine each track having a tiny sphere placed on it representing a planet. This is pretty much how the solar system works, with each small sphere travelling around the record and orbiting the Sun.

In reality, the planets are all spaced out at different distances, and these distances are vast. For example, let's imagine you built an imaginary road from the Earth to the Sun. If you got in your car and drove on that road toward the Sun at a constant 60mph, it would take you around 170 years to reach it. I hope there are rest stops along the way, as that's a long time to wait for the toilet! The outer planets are many times farther away, and some would take thousands of years to reach if we theoretically drove to them.

The ecliptic

Looking out from Earth, we see the orbital plane of the planets and the path of the Sun, Moon and planets. We call this the 'ecliptic'. The ecliptic is fixed in the night sky just as the constellations and Milky Way are, but where is it?

You may remember me mentioning the zodiac in previous chapters. The ecliptic – the path of the Sun, Moon and planets on the celestial sphere – is found in the zodiac. The zodiac is a collection of twelve constellations that correlate with the path of the Sun, Moon and planets. You may have heard of them before as 'star signs' or 'birth signs' in astrology – your zodiac or star sign is the part of the sky the Sun was in on the day you were born.

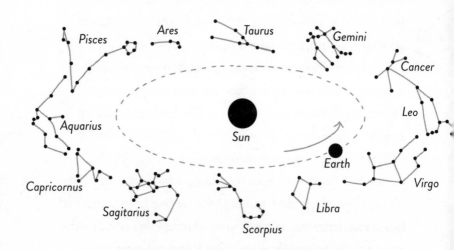

(Astrology is often confused with astronomy: astronomy is a natural science and the study of space and the night sky, astrology is a pseudoscience and the belief that the motion of celestial bodies influences our behaviour and lives.)

The zodiac is where the Sun, Moon and planets live and where you can find and identify them if they are visible at the time of looking. The twelve zodiacal constellations in order of appearance are Aries, Taurus, Gemini, Cancer, Leo, Virgo, Libra, Scorpius, Sagittarius, Capricorn, Aquarius and Pisces. (I'm an Aquarius if you were thinking of asking.) Not all of the zodiac are visible simultaneously, as we saw earlier. They also appear higher or lower in the sky, depending on the season.

The planets

The solar system planets do not stray from this path, but they move along it and occupy different positions at different times. An up-to-date sky guide or app will tell you what planets are on view at any one time and what constellation they are found in.

Some planets can be seen throughout the night, and these are the outer planets: Mars, Jupiter and Saturn. Uranus and Neptune require a telescope to been seen. The inner planets, Mercury and Venus, are always close to the Sun and are only visible around sunrise or sunset.

Occasionally, two or more planets appear to be grouped together and can even be grouped with the Moon. This is known as a 'conjunction' and is a beautiful sight. Sometimes all five of the visible planets, including the Moon, are visible in the same evening but close conjunctions with all of them are very rare indeed.

Let's find out more about the planets themselves.

Mercury

The planet Mercury, named after the Roman messenger god, is the smallest planet in the solar system and closest to the Sun. It isn't much larger than our Moon, and is a small rocky world devoid of atmosphere and pockmarked with craters, like the Moon. In the daytime, temperatures rise to a balmy 400° but they reach a chilly -180° at night. Mercury is inside the Earth's orbit and so is known as one of the two inferior planets, the second inferior planet being Venus.

Venus

Venus, named after the Roman god of love, is the second planet out from the Sun and is a similar size to the Earth. It's often regarded as Earth's twin, but size is where the similarity ends. Before modern astronomy and science, it was believed that Venus was a tropical world with jungles and rainforests. It wasn't until the early twentieth century that it was discovered to be a harsh world driven by volcanism and with a runaway greenhouse effect. The temperature on the surface of Venus is around 470° with crushing atmospheric pressure 90 times that at the surface of Earth – the same as if you travelled a kilometre down beneath the ocean on our planet. The unpleasantness doesn't end there; when it rains, which it does, it rains sulphuric acid. Russian and US space probes landed on the planet's surface, and even though they were built like tanks, they didn't last very long. It was only a few short hours before the planet destroyed them. A day on Venus is equal in length to 243 Earth days, plus the planet spins backwards. It's a pretty strange and unwelcoming place indeed, but a joy to behold in the night sky. Due to being shrouded in thick white clouds and its proximity to Earth, Venus is the brightest planet in the solar system as seen from Earth. It's often referred to as the Morning or Evening Star due to its proximity to the Sun and therefore only being visible at dusk or dawn.

The next planet in the solar system is Earth so let's move onto Mars – the Red Planet.

Mars

Mars is named after the Roman god of war and is about half of the size of Earth. The red planet is remarkably similar to Earth in some respects, with similar seasons, day length and conditions. The temperature can reach a comfortable high of 20° in the daytime but plummet to as low as -20° at night. The atmosphere is very thin at around 1% of the atmospheric volume of Earth; however, this is more than feasible for human bases and colonies, making Mars the perfect destination for interplanetary crewed space missions.

Mars is red due to the amount of iron and rust on its surface and it boasts the highest peak in the solar system with the extinct volcano Olympus Mons. Mars may also contain underground stores of water, which is tempting for future colonies. When looking at Mars through a telescope, its bright carbon dioxide ice caps are a joy to behold.

Past Mars is a region of space called the asteroid belt, which we will come back to later, and after that is the king of the planets, Jupiter.

Jupiter

Jupiter is named after the Roman king of the gods and is a massive gas giant so large you could fit the Earth inside it 1,300 times. One of its main features, the Great Red Spot, is a gigantic 300-year-old storm that could in itself easily swallow the planet Earth whole. Jupiter is a monster but a beautiful monster with unique weather systems and bands of multi-coloured clouds around it. The planet has no solid surface, just swirling layers of gas with a hot liquid metal or solid core.

Jupiter has many moons and is famous for its four Galilean moons: Io – the most volcanically active body in the solar system, Europa – with the smoothest surface in the solar system and believed to have a liquid ocean beneath its icy crust, Ganymede – the largest moon in the solar system, and Callisto – the most heavily cratered surface and third-largest moon in the solar system.

If you look at Jupiter through a telescope, you can see its

moons change position nightly as they perform their orbital dance around the planet. The moons and disc of the planet are visible with binoculars. A telescope will bring out more detail, such as the bands of cloud and Great Red Spot.

Saturn

We now move further out into the solar system, twice the distance again, to another massive gas giant – the ringed planet Saturn, jewel of the solar system. Saturn is a gas giant like Jupiter but slightly smaller in size. It is, however, less dense than water, meaning that if you could find an ocean large enough, it would float. Saturn's defining feature is its beautiful ring system made up of dust, rock and ice. The rings are massive at around 280,000km wide, but are only a few metres thick. They are believed to be the captured remnant of comets or a moon that broke up and was destroyed by an impact or Saturn's powerful tidal forces.

Like Jupiter, Saturn has many moons, with the largest being Titan, the second-largest moon in the solar system. Titan is also the only moon with a thick atmosphere, comprising mainly of nitrogen with a crust of solid ice and rock. Titan is extremely cold and has mountain ranges of pure rock and hard ice, along with seas, rivers and lakes of liquid methane.

Mars, Jupiter and Saturn are known as the superior planets, being outside of Earth's orbit.

Uranus and Neptune

The last two planets in the solar system aren't visible to the naked eye and can only be seen with telescopes or very powerful binoculars. These are the two ice giants, Uranus and Neptune, which are well worth hunting for if you want to delve further into astronomy.

The Moon

The planets aren't the only thing to appear to change position, of course. Our Moon does also, and to a much greater extent.

I'm often asked, 'Where is the Moon tonight?' or questions like, 'Why has the Moon changed position, and why does it look different to last week?' These are all valid questions and simple to answer. Like everything else in the night sky, the Moon is in motion, and because of its proximity to us, that movement is pronounced.

Before we go into the Moon's movement, it's a good idea to find out a little about it first.

The Moon is the Earth's only natural satellite. It is believed that it formed as a result of a collision between the early Earth and a Mars-sized planet known as Theia around 100 million years after the birth of the solar system 4.5 billion years ago. In Greek mythology, Theia was a Titaness who gave birth to the Moon. As Theia collided with the very young Earth, there was a cataclysmic explosion of material, and as both planets writhed, they were absorbed into each other. Not all the material was absorbed; though, much of it was thrown into space. After another 100 million years, this material coalesced gravitationally into the object we now know as the Moon. It was much closer 4.5 billion years ago but has been, and still is, moving away from the Earth at a rate of 4cm each year. Right now, it is ideally placed for stunning solar eclipses when it occasionally passes between the Earth and the Sun. In the future, though, these will be much less spectacular as the Moon will no longer completely block the Sun as it passes in front of it.

The Moon is roughly the same diameter as the continent of Australia, at 3,400km. It is a sphere, as is the Earth, but it is covered with the historical evidence of its tumultuous past, with craters and the evidence of massive lava flows that are now referred to as the Moon's seas or, in Latin, 'Mare' – these are the dark patches we see as we gaze upon it. There are valleys and mountain ranges, as well as many other topographical features. However, it is utterly devoid of any atmosphere due to a weak magnetic field and its small size. At the time of writing this book, it is the only solar system object away from Earth that humans have set foot on with the Apollo missions. Yes, we did go to the Moon and no, the Earth isn't flat.

The Moon is large enough to affect Earth though, and its gravity tugs on the Earth's surface, creating the rise and fall of tides in our seas and oceans. When it's high tide, the

Moon is pulling at the ocean and 'lifting' it. When it is low tide, the Moon is on the other side of the planet, and its gravitational effects are not as strong.

The Moon orbits the Earth approximately once a month, 'month' meaning 'one Moon'. During this orbit, it can be seen along the ecliptic, the path mentioned above. Unlike stars, the Moon does not emit its own light, it only reflects the Sun's light, and so we see the parts of the Moon that are illuminated or not. Early in its lunar cycle we see a New Moon when the Moon is invisible due to being placed between the Earth and the Sun, with the Sun's light shining on the side pointing away from us, and the side of the Moon facing us in darkness. As the Moon moves around in its orbit, it is illuminated more and more, creating the lunar phases.

This is first noticeable near the Sun in the west after sunset. We often spot a very thin crescent Moon. As the month moves on, it moves higher and further east as it moves along the invisible line of the ecliptic in the zodiac. Its cycle reaches phases such as First Quarter, Full Moon, Last Quarter and back to New Moon, always in a west to east direction along the ecliptic. Each Moon phase usually lasts a day or two before it can be seen to change phase and position in the sky again.

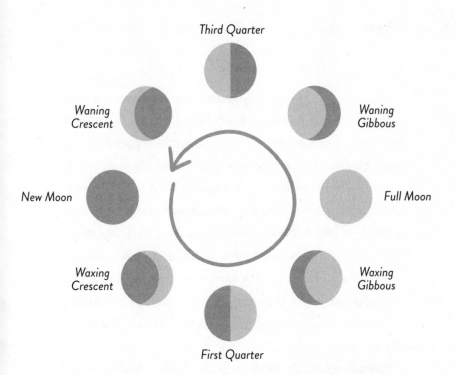

The Moon is one of the most beautiful objects in the night sky and the easiest to spot. An early crescent moon is stunning, especially if the part not in direct sunlight is glowing from the light of the Earth reflecting off it. This is called Earth Shine. As the lunar cycle continues through the month, more and more of the Moon is illuminated, culminating in the Full Moon, when its entire disc reflects the sunlight directly back at us from the Sun, which at this time is behind the Earth.

The Full Moon is a monthly occurrence, and it falls on different days due to the Moon's orbit not being precisely one month long (taking roughly 29½ days to complete).

Moon magic

If you have been outside during a Full Moon, you will know how bright this is. All but the brightest night sky objects are drowned out by the light from the Moon, and you can easily navigate your surroundings by its light, and even see on some occasions within its shadows.

The Moon's orbit isn't a perfect circle but is slightly oval, as are many of the orbits of other objects in the solar system. When the Moon is Full and close in its orbit to Earth, it can appear slightly larger and much brighter than usual. This is now referred to more and more as a Supermoon; some traditionalists get grumpy about this, but it encourages people to look up, which is a thumbs up from me.

Standing outside in the light of a Full Moon is indeed quite magical, with the eerie but pleasing bluish-white light it produces; it always makes me feel happy. Make sure you look around you and experience the light of the Moon as it touches the natural world. Since the dawn of man, we have been bathed in its calming and purifying light. It has been an integral part of civilisations, societies, religions and farming communities, and is steeped in folklore. You will hear people talk of a Harvest Moon, Wolf Moon or Hunter's Moon, as well as other names. These are the titles given to each Full Moon by Europeans and Native Americans after features they associated with the seasons. You may know the phrase 'once in a Blue Moon', meaning a long period of time or rare occurrence. This is correct as a Blue Moon is very rare; it can be either the third Full Moon of an astronomical season, such as the third Full Moon in summer, or the second Full Moon in a calendar month, which is possible if the Moon is Full at the start and then again at the end of a month longer than 29½ days.

Another name you may hear is a Blood Moon – not to be confused with a reddish-orange Moon seen low on the horizon through a thick dusty atmosphere. You will notice as the Moon gets higher in the sky, it gets whiter. This is because we are viewing it through a thinner atmosphere and less dust, making the light whiter. The discolouration of the Moon at sunrise and sunset plus the Moon also looking larger near the horizon is just an optical illusion. A Blood Moon refers to a total eclipse of the Moon, where the Earth passes in between the Moon and the Sun, and the sunlight passes through the thick atmosphere of the Earth, turning the Moon an orange red.

Studying the Moon instils a calming effect, not just with the naked eye but also through binoculars or telescopes, where its amazing topography and detail come to life. The wrong time to look at the Moon up close is when it is Full. Then, most of its features appear flat and washed out due to the lack of shadows, and you can even damage your eyes with its powerful glare, so always use a moon filter attachment, which adjusts the brightness of the Full Moon when seen through a telescope.

If you look at the Moon with binoculars or a telescope

when it's not a Full Moon, when its phase is at First Quarter perhaps, you'll see that all of the features running down the 'terminator' (the dividing line where daylight turns into night-time) are displayed in stunning detail, with long shadows highlighting the topography. A good Moon atlas or map will help you explore this amazing world further.

Next, we look at some other objects that can appear in the night sky – some of the most dramatic and breathtaking sights a stargazer will see.

Chapter Eleven

SPACE ROCKS

The night sky, I think you will agree, is full of wonderful things: stars, constellations, galaxies, star clusters, nebulae and more. The Milky Way stretches from horizon to horizon, waiting for the next gasp of joy as someone spots it for the first time, and the Moon and planets wander along the ecliptic in their celestial dance around the Sun.

Imagine you are sitting out in your garden looking up as you tune in to the nocturnal world around you, with your senses sharpening at the things that go on in the darkness. You are calm, and everything is peaceful and tranquil. You keep your gaze fixed on the sky as you take a sip of hot tea or soup. Everything is perfect, and you are calm and well. The night sky is chilling you out, and it is fantastic. Then, all of a sudden . . . Whoooooosh! A bright fiery streak of light enters your field of vision and shoots across the sky.

It lasts as long as the forced breath you suck in. Some

people leap up and down doing a little victory dance, while others recount the brief moment they just experienced like awestruck children. No matter what your response to this kind of thing, you have been lucky to spot a meteor, otherwise known as a shooting star, which are indeed nature's fireworks.

'Space weather'

Look up for an extended amount of time, and you will see meteors, I guarantee. It might take an hour, or it may only take a second. Some of the best meteors seen are spotted by people going about their nightly business – putting the bins out or getting the cat in. Meteors appear randomly in any part of the sky at any time; most evenings will see one or two faint or small ones that appear and burn up in the blink of an eye. Very occasionally, there will be a bright fireball that can last many seconds. More and more of these are being caught on household CCTV and video doorbells, as well as more people reporting them on social media. Don't worry, the War of the Worlds hasn't started; technology has just improved, and more of us are now our very own news channel.

Meteors are part of a collection or grouping of events and objects called 'space weather'.

Space weather involves something hitting and then burning up in the Earth's atmosphere, such as meteors, or the charged particles from the Sun that cause the northern and southern lights (known as Aurora) at or near the Earth's poles. If you ever get the opportunity to go somewhere like Norway or Iceland to view Aurora, take it. Most who see it are blown away by its beauty and splendour. It is, as they say, a bucket list moment.

In this chapter, we concentrate on the things we can see from any part of the planet, and these are space rocks.

Meteors and shooting stars

First, let's go back to our friend, the shooting star, or meteor to be precise. Meteors are just space rocks moving around the solar system in various directions, and they have been doing so since the solar system was created. They aren't called meteors until they encounter our atmosphere, or that of another planet. While they are floating around in space, they are known as meteoroids. These can range in size from a grain of sand or dust to the size of pebbles, small boulders or even bigger and are often referred to as 'space dust'. (Large pieces are called asteroids and can be massive – see below.)

Smaller meteoroids encounter the Earth or another planet's atmosphere as it ploughs into them or they plough into it, and turn into meteors. As soon as the meteoroid meets the atmosphere, the brakes are slammed on, and it begins to heat up very quickly through the friction caused by the drag of the atmosphere. Smaller grain-sized meteors heat up and are annihilated as a bright flash or streak of light in the blink of an eye, but larger rocks can hold their own and last a lot longer as they burn up and they end as a bright fireball streaking across the sky. These brilliant fireball events can last several seconds and even hit the ground if the meteor is large enough. This occasionally happens, and when they make it to the ground they are called meteorites. It is rare for meteors to make it down, though, with most ending their journeys as a bright flash of light as they burn up high in the atmosphere.

They contain different elements and can display bright colours as they burn up in the atmosphere: green, yellow or red, and some can even be blue. It all depends on what they are made of, and the different elements they contain – sodium can burn yellow, or magnesium for green, etc.

Meteor showers

Meteors happens at random and can't be predicted; however, there are times when you can expect heightened meteor activity. These are called meteor showers. At certain times of the year, named meteor showers occur where the rates of meteors that can be spotted are greater than usual and, in some cases, far more significant, with predicted hourly rates reaching or exceeding many tens of meteors per hour. This is undoubtedly the case with the Perseid

Meteor Shower in August and Geminid Meteor Shower in December.

A meteor shower gets its name from the constellation it appears to emanate or radiate from. This point is called the 'radiant' so, for example, the radiant for the Perseid Meteor Shower can be found in the constellation of Perseus, and the Geminid Meteor Shower in the constellation of Gemini.

There are over a dozen well-known or annual meteor showers a year, and a good sky guide or app will tell you when they are most active. I would like to mention that when you hear about a meteor shower that is due, the magazines, books and news channels focus on the radiant and point people toward it as a place to look for the meteors or sit there in ambush like a hungry animal. This is the direction the meteors will be coming from, not where they can all be seen.

Meteors will appear randomly in any part of the sky. If you trace their path back to the meteor shower's radiant, they are referred to individually as a Perseid, Geminid or Lyrid, depending on the shower they come from. Some meteors, however, will appear to have come from a different direction entirely, and these are 'sporadic meteors' – rogue background meteors. Just as remarkable to watch, though!

It is rare for meteor showers to include meteors much larger than grains of sand or pebbles due to the origin of meteor showers. Unlike background meteor activity from meteoroids randomly moving around in space, meteor showers originate from the tails of comets. More precisely, when a comet (which we will cover shortly) passes near the Sun, it releases dust, gas and gravel into its tail, eventually leaving streams of debris and stardust around the solar system. The Earth encounters some of these debris streams in its orbit, and as it ploughs into them, the results are meteor showers.

Meteor showers vary in their intensity, from shower to shower and year to year. Most feature a meteor or two each minute, but others have been known to 'outburst', when hundreds of thousands of meteors rain down. This is extremely rare, but the Leonid meteor shower has outburst a number of times over the last few hundred years, with a regular outburst every 33 years. The most noteworthy were in 1833 and 1966 when hundreds of thousands of meteors fell like rain, amplifying the feeling of the Earth's movement through space. Frightened witnesses held on to objects on the ground as the feeling of Earth speeding through space was incredibly strong. The shooting stars came from a point in the sky in the constellation of Leo, the Leonid radiant. The Leonid meteor shower's parent comet is comet Swift-Tuttle, which orbits the Sun every 33 years and on occasion, as it sublimates with the heat of the Sun, releases tremendous amounts of material.

The people witnessing these major events didn't have to plan for or go out of their way to see the meteors as there were so many. Usually, though, it's a case of blink and you miss it with meteor showers, so it's a good idea to be prepared.

With meteor showers, you tend to be outside for quite some time, so dress warmly and comfortably. Your observation spot needs to be thought out well in advance also. A spot away from direct sources of light is advisable, and objects blocking your view, such as buildings and trees, need to be avoided as much as possible. You want to be able to fill your gaze with as much sky as possible. The middle of a field in the darkest depths of the countryside is as perfect as it gets but, alas, many of us don't have the luxury of living near such skygazing spots. The other important thing is to be comfortable, so lie down on a reclining chair, picnic blanket on the ground, or even on a trampoline. Garden trampolines are perfect for stargazing; just make sure you have plenty of blankets or sleeping bags to cover yourself with.

Meteor spotting is fun, so make an evening of it. You can find out when meteor showers are due in sky guides or apps, but everyone loves the Perseid Meteor Shower, which can be seen from 17 July to 24 August and peaks around the 12 and 13 of August. This is one of my favourites as it is during the warmth of summer. The meteors that the Perseids produce are simply stunning, with fast, bright fireballs being common. You will need a clear sky, though, and meteor showers don't always perform. I've spent many an hour disappointed with a lack of meteors or with a sky shrouded in cloud. This is the random nature of meteors and meteor showers, which makes them quite exciting as you never know what you will get. You just have to go out and see.

Asteroids

Meteors aren't usually dangerous and only become dangerous when they are large. There are no reports of humans killed by meteors, but recently many were injured by an object that exploded in the atmosphere near the town of Chelyabinsk in Russia, one of the most notorious meteor events of the twenty-first century. A small rock didn't cause the meteor in this event; it was caused by an asteroid that was 20m in diameter and weighed 10,000 tons. The force and the heat generated as it ploughed through the atmosphere at 40,000mph caused it to explode with the energy of a nuclear weapon about 18 miles above the surface of the Russian countryside. Windows and doors were blown out, and people knocked to the floor suffered injuries from falling debris and suffered from temporary flash blindness.

This object was an asteroid, and not even a big one at around 20m. Some can be many tens or even hundreds of miles in diameter.

An asteroid is a large piece of space rock left over from the formation of the solar system 4.5 billion years ago. They tend to be found in the asteroid belt between the orbits of Mars and Jupiter. Space agencies such as NASA and ESA keep an eye out for any rogue asteroids, which could pose problems for our planet. None are forecasted to impact, but some do get quite close.

Comets

Asteroids and comets are the staple diet of many disaster movies and have been responsible for events in the past. Comets are similar to asteroids in size and shape but comprise dust, rock and ice. They originate from the furthest parts of the solar system, known as the Oort Cloud and Kuiper Belt.

Comets are the remnants of when the solar system was created. Some of these massive icy bodies intersect with each other and are knocked into the inner solar system, where they can establish wild, eccentric orbits around the Sun or plummet into it. As a comet draws closer to the inner solar system and the Sun, it warms up, and the ice, dust and gas sublimates at its surface, ejecting it into space; the ejecta can form massive comet tails many millions of miles long. These can appear to hang in the night sky for many weeks if they become visible and are beautiful objects to observe.

Many people understandably get confused between comets, meteors and shooting stars. Comets appear in the sky and linger, not noticeably moving for many weeks. Shooting stars or meteors can appear and then disappear instantly, or last up to several seconds before they either burn up or impact the ground.

As with asteroids, comets can be massive and many tens of miles across. None are known to be a threat to Earth in the near future, but asteroids and comets have been responsible for causing havoc and, in some instances, almost wiping out all life on Earth in the past. The most famous theory is that a comet or asteroid several miles in diameter was responsible for wiping out the dinosaurs after slamming into the Yucatan Peninsula in Mexico 66 million years ago. The impact caused catastrophic devastation at the moment of impact, which continued as a thick shroud of dust enveloping the planet, killing most life on Earth. Plants, large dinosaurs and other animals expired. The few survivors were much smaller, and among these were small mammals. Over a very short period, the dinosaurs' reign was over, leaving the door open for mammals to become the dominant species on the planet.

This isn't going to happen again soon, but NASA and other space agencies are looking at ways to deflect these large objects if ever one was deemed too close for comfort, so there is no need to worry.

• • •

So, we now know that space rocks can range from tiny space dust and meteoroids to mountain-sized asteroids and comets. Asteroids tend to be invisible to the naked eye as they are deeper, darker solar system objects, visible as faint dots changing position from night to night in images.

If they are large enough and close enough to the inner solar system to form a tail, comets are a rare and very beautiful sight, looking frozen in time. Meteors are the race cars of the night sky, impacting our atmosphere at tremendous speeds and ending their journeys as bright flashes of light or fireballs that sometimes make it to the ground as meteorites.

It doesn't stop here, though; there is still more to see in the night sky.

Chapter Twelve

SPACESHIPS AND THINGS THAT MOVE IN THE SKY

So far, this book has been about the natural phenomena in the night sky above. As you have learned, the night sky is a rich and varied place.

As you look up perhaps briefly, or for a more prolonged stargazing session, you will see the familiar stars and constellations for the time of year. Sometimes, as you scan the heavens, something new catches your eye. Not a meteor or shooting star, but something different. It's a star-like object, but it's faint and moving. It doesn't have the usual flashing green or red lights of an aircraft and it is silent. It's odd and feels like it shouldn't be there – an interloper.

Perhaps you note a faint flash or succession of light flashes in the sky. Or you spot what appears to be a slow-moving star then, all of a sudden, it gets brighter and flares up before dimming again and disappearing. Sometimes things appear and disappear. Some moving lights can even be seen passing over slowly in formation – in pairs, a triangular shape, or even a moving constellation of multiple objects.

This could be quite alarming for someone who sees these for the first time and has a wild imagination and an excitable mind. A UFO, possibly? To the uninitiated and those subscribed to pop culture, definitely. In actual fact, what you have spotted is an artificial or human-made military, scientific or communication satellite, or a number of them.

In addition to these, there are pieces of space junk, dead or disused satellites and the left-over components of past rocket launches lingering in our night sky.

Space junk

On any clear night with good conditions, you can see many of these objects passing overhead in all different directions. An hour or two on either side of sunrise or sunset brings many. The summer is also a good time due to the Sun not being far below the horizon at this time of year. They can be seen all year round and at any time of night, and even in daylight if you are fortunate, but this is rare. The satellites and space junk overhead reflect sunlight from their polished surfaces back toward Earth, and this is how we can spot them. They range in size from a toaster to a school bus, and there are many thousands of them in various orbits and altitudes, with over half being space junk left behind from satellites that have outlived their purpose or are dead in space, some being whole rocket stages. They criss-cross the sky in all directions, and you can sometimes see several at once, with the sky appearing alive or crawling with them, as some might say.

Many spotted these tiny objects during the Coronavirus lockdown of 2020, and people enjoyed looking for them. As well as the communications satellites and old rocket bodies, trains of many tens or more of Starlink internet satellites were witnessed traversing the sky numerous times. Each satellite followed one after the other as they jostled for their final orbital placements, looking like a string of pearls flying across the sky, becoming more spread out as the weeks and months went by.

They can be fascinating to watch, and it feels like a new dawn of space exploration is ready to burst forth with all this new technology as we venture back to the Moon and onto Mars.

Unlike meteors, which occur at random, satellites and most space junk are catalogued and can be predicted. Some websites and apps do just this, and also identify objects that have already passed over. You can take your satellite observing to the next level and get super-geeky about it, but for now, we will move on to something much, much bigger . . .

The International Space Station

The International Space Station – ISS – is an orbital take-over and habitat. It is the largest human-made object in space, with a similar footprint to a football pitch and the same amount of space inside as a large passenger jet. It travels at 17,500mph around 200 miles above the Earth's surface while orbiting the planet approximately every 90 minutes, so approximately 15 times a day. There's usually an international crew of three or six, but recently that has increased to nine on occasion. The ISS has been in orbit since 1998, initially starting with the first Russian module and slowly growing to its final configuration over 20 plus years.

On board the crew carry out numerous scientific experiments in microgravity and monitor the effects of low gravity on humans and other organisms for future missions in space. The crew rotation is usually six months to a year in space before being replaced. Crew transport and supplies are sent by Russian, American, European and Japanese rockets and spacecraft. More and more commercial spacecraft now take over from where the older state-run shuttles and Russian re-supply and re-crew flights left off. The ISS is expected to be in service until the 2030s, and you can see it just using your eyes. I'm hoping it remains in operation for much longer as it is a fantastic sight to watch overhead and a beacon of human achievement.

About every six weeks or so, conditions are right for

visible passes over any one location. This is down to the orbit of the Space Station. Passes happen around sunrise and sunset, with multiple passes being possible throughout the evening, the closer to the summer solstice you are.

Sometimes passes can be seen throughout an entire night due to the Sun not being far beneath the horizon. All you need to know is when to look and what direction to look in. This information can easily be obtained via apps or the internet, and I even do alerts and countdowns for the UK on my social media channels.

Once you have found the correct time and pass details, you don't need any equipment to look for the ISS. It is one of those times where you can watch in shorts and flip flops or even your pyjamas. The International Space Station typically passes over in a westerly to easterly direction and it lasts around six minutes at the longest.

Watch in the direction specified at the right time; it may take a little while to spot as the passes usually start lower in the murky part of the sky, and the ISS is less bright. Once you have managed to spot it, you will notice it getting higher and brighter as it approaches, looking like a bright star moving across the sky. This is especially so with a pass that is forecast to be overhead. It can start anywhere in the sky, depending on when you are looking and your particular location; then, as the ISS passes over, it can be the brightest object in the night sky except for the Moon, and seen even from bright light-polluted cities such as London. Once overhead, remember to wave at the astronauts and send your best wishes; they might be doing the same back.

Once you've dispensed with the pleasantries, the ISS will head toward the eastern horizon, sometimes fading out before it gets to the horizon as it enters Earth's shadow and is no longer illuminated by sunlight. The ISS has no noticeable self-illumination, but because it is so large, it reflects a lot back to us. Occasionally, re-supply and crew spacecraft can be seen travelling near the Space Station in front or behind; these appear as smaller fainter objects and are proper spaceships, flying in the sky over your house!

Tomorrow's world

This is where the magic of it all kicks in with many people, as they watch the future of humanity fly over their homes from their gardens or doorsteps. Adults, children, the young and the old are fascinated, awestruck and inspired. I'm not sure if this was intentional when the ISS was first planned, but it has undoubtedly encouraged many people to look up.

But it doesn't stop there; more and more rockets and spacecraft launches, crewed and uncrewed, are streamed live, most of which are launched from the east coast of America at locations such as Kennedy Space Center. This is especially good for people in the UK and other parts of Europe as many launches have their trajectories pass over this part of the world, and if the timing is right and the

.weather clear, you can watch a launch live on the internet and then see the spacecraft fly over your house around 15 minutes later.

The sheer amount of excitement, joy and satisfaction this gives to everyday people is genuinely profound. You don't just get to experience space travel through a screen; you can see it in all its real-world glory just with your eyes as it flies over you. You don't forget moments like these. Many take pictures of the ISS and other spacecraft passing over with cameras or even their phones, and some of the results are truly breath-taking. (We look at night sky photography later.)

We aren't just looking at the aeons of history and the beauty of the night sky as we look up anymore; we now also see the future. A future where we return to the Moon and on to Mars and beyond; a future where space travel will eventually become as easy and as regular as air travel, as we venture out further establishing colonies and homes in our solar system, and eventually interstellar space. The stuff dreams are made of will become more and more accessible to you and me. I can only imagine the adventures and wonders that await us.

For now, we mere mortals will have to make do with the beautiful night sky as it is from Earth, as a select few blaze a trail for the rest of us. We can, however, continue to extend our exploration of the night sky, and for this, you will need binoculars and a telescope.

Chapter Thirteen

GADGETS – TAKING YOUR STARGAZING A STEP FURTHER

One of the first images that springs to people's minds when you mention you enjoy stargazing or astronomy is the stereotypical image of someone looking through the end of a telescope. Indeed, it is a rite of passage to many and a crucial item, as is a fishing rod to a fisher, or an easel and brush to a painter. Some believe it is a must-have to make them a proper astronomer but you don't need a telescope at all if you are new to stargazing, or are just a casual observer. However, they are nice to have.

In this chapter, we will look at the gadgets you can use to take your stargazing and astronomy further, and to delve into the world of astrophotography. We have learned the basics of the night sky and how to find our way around it. With practice, we can look up and identify different things quickly and easily, and have a better understanding of what is up there. What we see with the naked eye is just scratching the surface, and there is a whole universe to explore more deeply with the aid of some traditional astronomy equipment.

Some reading this book may have a telescope already, and there may be others who have been impatiently reading this book hoping to get to the good part about telescopes and all the fantastic astronomy kit you can buy. But read the rest of this book first! It is essential for you to get the most out of your stargazing, and you will find it immensely helpful in enjoying a new telescope more. You must learn to walk before you run, as the old saying goes.

When people first develop an interest in the night sky, they tend to go to the internet, and may get some books or magazines on the subject filled with mentions of telescopic objects and the wonders to see, alongside glossy adverts for that latest state-of-the-art telescope. You read the reviews, and make your choice. It seems helpful that your purchase means that you don't need to learn about the night sky as it has a built-in computer with hundreds of thousands of objects stored in its memory, all ready at the push of a button for the telescope to find and you to view. All you need to do is set it up, and it will locate all the wonders of the night sky for you. Deep joy! So, you order one. (I fell into this trap too.)

Technology is great when it works and operates as

advertised, but you can easily get sucked into a trap with telescopes. Some amateur astronomers will pay the same amount of money for a telescope as they would for a car. The problem with small amateur computerised telescopes, or what are called 'GoTo' telescopes, is their build quality, the computer, and the size and quality of the optics. Optics are expensive to produce and are the heart of a telescope. If you want a computer and all the toys, the optics will be small unless you are prepared to spend a lot of money. The larger or better quality the optics are, the brighter and clearer the image will be. Magnification, however, is governed not just by the telescope but by the eyepieces, the essential part where quality matters. Some high-quality eyepieces can cost more than the telescope itself but are more than worth it. A poor telescope with high-quality eyepieces will give great images. A high-quality or large telescope coupled with excellent quality eyepieces will be a joy to use.

One other thing to avoid is cheap department store or supermarket telescopes. They are usually small with poor quality optics, which don't give good views, and can be cheaply made. In some instances, they aren't much better than toys. If your budget is small, don't get a cheap telescope; you will be simply wasting your money and will be disappointed.

So, how do you find a telescope that is easy to use, isn't as expensive as a computerised model, and gives excellent views? I recommend keeping it simple, and using the right telescope for the job. And before we go on, my advice to anyone thinking about getting their first telescope is to purchase a pair of good quality binoculars first.

The right telescope for the right job

As with most things, there are telescopes for different jobs. Some star followers have more than one telescope, each being a specialist in its own right. I must point out that you are never going to get the views you see in Hubble images in certain magazines or websites from a telescope. The human eye isn't capable enough. Objects seen through telescopes with the naked eye are usually very faint or small. The better the optics, the brighter and crisper the views will be, but even still, only photography can produce images in detail or close up, and there is a lot of work involved to create them (as we look at later in the chapter).

There are two main parts of a telescope: the optical tube assembly, the bit that produces the image, and the mount, which is what the optical tube assembly sits on.

Mounts

A tripod allowing the telescope to be moved up and down and left and right is the most basic form of mount. The Altazimuth mount allows you to simply move the telescope tube up and down or left to right either by hand or by computer.

You can add an 'equatorial head', which allows you to align the telescope with the polar axis of Earth and so can track the motion of stars in the sky with levers and slow-motion controls. This is the most common type of telescope mount, but it can be a steep learning curve and quite complicated.

In addition to this, there are fully computerised equatorial mounts known as 'GoTo'. Still, these are much more expensive and more useful for astrophotographers who want to track an object for an extended period precisely.

Telescope optics

The first type of telescope optical tube assembly is the refractor telescope. These telescopes are your typical telescopes, or what comes to mind as the traditional telescope on a tripod where you look through the thinner end. Refractors use a lens known as a primary lens at one end of a tube with a smaller lens or eyepiece that you look through. Refractors tend to have smaller optics, up to four inches in diameter, and are suitable for brighter objects, especially the Moon and planets.

The next type of telescope optics tube assembly is the Newtonian Reflector Telescope, invented by Sir Isaac Newton. Newtonian telescopes use mirrors instead of lenses. These is cheaper to manufacture and can include huge mirrors, on average eight to ten inches in diameter going up to twenty inches. The large mirror, known as the primary, sits at the base of the optical tube and reflects light up the tube to a secondary mirror, reflecting the light into the eyepiece on the side of the tube. Newtonians can usually be found on top of equatorial mounts and are giant light buckets and capable all-rounders due to their tremendous light-gathering ability.

Following on from these, we move onto the more advanced and more complicated optics of Cassegrains, where a combination of lenses and mirrors are used. These are the domain of the advanced amateur and astrophotographer, ranging in sizes upward from eight inches on average, and are usually perched on either computerised Altazimuth or equatorial mounts. They are complex and expensive, giving near-perfect views.

There are always hybrids and mash-ups and new technologies, but the above is the mainstay of any telescope store. The problem is, we still haven't found that perfect telescope for the beginner and advanced astronomer alike. All of the above have caveats, so we need a telescope that only has a few, and for that we look at what is called a Dobsonian Reflector Telescope.

Dobsonian Reflector Telescope

Dobsonian telescopes were invented by John Dobson, the co-founder of the Sidewalk Astronomy movement in San Francisco, California in the 1970s. The telescopes they used were cheaply made from plywood, cardboard and the glass from old ship portholes from the nearby dockyards. Using the same design as a Newtonian optical tube assembly, the glass portholes were hand-ground and the cardboard or other material tube assemblies were placed on a very simple plywood Altazimuth mount. These home-made telescopes were simply brilliant and helped spawn a massive astronomy movement in America.

The design has been refined with modern manufacturing and engineering techniques to produce one of the best telescope types available – huge light-gathering optics that capture the faintest objects combined with simple wood or metal mounts known as 'rocker boxes'. The benefit of these telescopes is that anyone can set them up and use them. There is no set-up time, no alignment, no complex assembly, and no computer to have to faff with. You sit the tube on top of the rocker box and away you go; you are stargazing, and the views are spectacular. The only downside is that you can't take long exposure images as the telescope doesn't track the sky, and you need to have an idea of what you want to see.

I always recommend Dobsonian telescopes to beginners or infrequent users as you can just get on with it and enjoy the night sky with the huge optics. When it comes to taking pictures, though, the Dobsonian needs to step aside as we use the other telescopes mentioned above to take photographs.

Binoculars

As well as a telescope, binoculars are a must for every star-gazer or budding astronomer. I use binoculars more as they are easier to use and I keep a pair near my back door and a pair in the car. They are great for quickly grabbing when you want to look at something without carting telescopes outside.

They come in various sizes, but the best size for astronomy is 10x50 or larger. I use 10x50, and they are great all-rounders for day and night viewing and are small and light. However, when it comes to stargazing, I go much bigger with a pair of 15x70 or 20x80. You can get huge 25x100 or even larger binoculars that are two short refractor telescopes strapped together using telescope eyepieces, but these are very expensive.

Usually, binoculars sub 20x80 will cost you less than a top-quality pair of running shoes or trainers and far less than the price of a telescope and are as much if not more fun. Binoculars are simple in themselves. All that is necessary to start is to make sure the dioptre adjustment, usually on the right eyepiece, is set so the image in both eyes is sharp, and you only have to do this once. I focus the left eye first with my right eye closed, and then close my left eye as I fine-tune the focus of my right eye with the dioptre. Voila! Perfect views.

The only issue you will experience with binoculars, especially with large pairs, is their weight and the onset of fatigue, which can prevent you from holding them comfortably or steady. The good news is you can attach most binoculars to photographic tripods using a basic attachment, and then you can enjoy an effortless hour of use with your binoculars.

The world of binoculars and telescopes can become very technical and confusing, especially to beginners. In this chapter, we have covered the basics, which are enough to provide you with a more informed decision if you choose to purchase either or both.

With the right equipment for you, you will enjoy stargazing more, and it will be more satisfying and relaxing. Make it hard, and you won't be doing it for long. An hour spent at the eyepiece is just as thrilling and relaxing as if you were just looking up. It's not about the binoculars or the telescope; it's about the views they produce and how they make you feel.

Next, we will look at how to take pictures of the night sky and save those views.

Taking pictures

Not only can you enjoy the night sky looking up, or observing through a telescope or even binoculars, you can take pictures of the night sky too. Astrophotography or astro imaging is becoming more accessible to those who fancy taking photographs of the heavens, of beautiful sweeping Milky Way vistas or long exposures of the International Space Station as it passes over. You could capture an entire evening's worth of shooting stars from a meteor shower, and delve deeper into the cosmos capturing images of the Moon and planets, or even the universe and galaxies, nebulae and clusters.

Unlike your holiday snaps, taking pictures of objects in the night sky requires careful planning and the right equipment. It's worth saying that most things are very faint and don't want to say cheese, making them tricky to capture. As a result, it can turn into a very rewarding, satisfying hobby in itself and can be quite time consuming if you are a perfectionist slaving away for that perfect Andromeda image. As with everything, there are different levels of equipment and commitment required, but much of it is accessible to novice stargazers.

Astrophotography was the domain of pipe-smoking astronomers wearing tweed suits or lab coats in the olden days, with access to professional observatories and specialist photographic equipment. Even with all that tweed, equipment and pipe smoke, the images produced were more technical than appealing to the eye. Things improved when large ground-based observatories and the Hubble Space Telescope shared amazing views of objects deep within the heavens. The problem was that taking pictures of the night sky and its deepest secrets was still out of reach of ordinary people. Then came digital cameras, DSLRs, webcams and smartphones; the day of digital astrophotography had properly dawned.

Smartphone photography

You no longer need to be part of an academic institution or a professional astronomer to take images of the night sky, you just need to know how and to have the right kit. If you have a smartphone, you can start straight away. All you need is to make sure your phone can take long exposure images. Some have this feature built in, and some require an app with this feature.

With a smartphone, you can take long exposure images of star trails showing the motion of the sky, the Space Station and satellites, or even meteor showers. If you own a telescope, it's easy to attach your phone to the eyepiece with a low-cost adapter and capture images of the Moon and planets. Smartphones are a great way to take pictures of the night sky, but they are limited. However, as technology improves, they are becoming more and more capable.

DSLR/other cameras

The next step up from a smartphone would be a DSLR camera or any camera that allows you to manually control the camera's settings such as exposure, aperture, ISO, etc. DSLRs are the Swiss army knife of cameras and are versatile, with the ability to interchange lenses, tailor settings and even connect directly to a telescope if you have one. There are many controls to help you dial in that perfect exposure, and the great thing with digital photography is you can experiment to your heart's content.

Experimentation is the key to great images. I often see

a stand-out picture on social media and share it, which usually brings on a lot of praise and questions about what settings and equipment were used to get the results and the steps taken to obtain the image. This can be helpful; however, equipment will be different as well as conditions so I always advise using other people's methods as a guide only and to experiment with your own equipment. This way, instead of just copying settings, you will make mistakes and get varying levels of results from bad to good, which means you will learn and understand the process yourself. The same goes for all kinds of astrophotography, from simple smartphone shots to advanced imaging with a telescope. Trial and error and experimentation will always help you as well as fit better with varying conditions and situations.

DSLRs are good at capturing wide-field vistas of the Milky Way if you use a wide-angle lens. They can also be fantastic for the Moon and even some deep sky objects with the right lenses, though these can be hideously expensive. A DSLR comes into its own when attached to a telescope, and the list of objects you can image becomes almost endless. It's not as straightforward or as simple as pointing your camera at the night sky or an object through a telescope and just taking the picture. The night sky is a dark place and, as you would soon find out by trying to take a standard shot of it, there isn't a lot of light there, so your image will be very dark. This is where long exposures, stacking video frames or images and other methods come into play: the camera shutter is left open long enough to let in enough light for the image to become clear without overexposing it. You can then stack multiple exposures of the same image for even more detail in dedicated astro imaging software or apps like Photoshop.

The problem with long exposures is, of course, the apparent movement of the night sky. As the Earth spins, the sky moves, and this is very noticeable in long exposure night sky photography, with even the shortest exposures showing star trails. Star trails can be beautiful but, for most things, you will want stars and other objects to be pinpoint sharp and not display any movement. For this, you will need a tracking mount – a device that tracks or follows the stars, keeping the desired object stationary in the camera's or eyepiece's field of view. These can be small units specifically for cameras to be attached to or built into a telescope's mount, as discussed earlier. These tracking systems can be as advanced as a top-end GoTo mount or as simple as a clock drive attached to a mount that enables it to follow the motion of the night sky, and they are required for much astrophotography.

Astrophotography doesn't have to be that involved and can be done with a telescope that doesn't have a tracking mount, such as a Dobsonian telescope. This is especially the case with imaging the Moon, planets and up-close images of the International Space Station. All you need is a simple video camera, such as a webcam in its most basic form, attached to the telescope at the eyepiece or in place of an eyepiece. You can then guide the telescope by hand and keep your subject in view, as the camera records video of the Moon, a planet or ISS as it passes over. Once captured, each video frame is stacked in specialist software, removing bad frames and producing a sharp image at the end of the process. Seeing detail on the ISS or feeling like you could reach into the picture and touch the Moon can be exhilarating, not to mention the hits you will get on social media!

If you continue with astrophotography and want it to be a serious hobby or interest, you will probably want the kit to go with it. DSLRs have their limits, as do telescope optics and mounts. Serious astrophotography enthusiasts will have top-of-the-line mounts and telescope optics coupled with state-of-the-art CCD cameras or other technologies inside purpose-made astro imaging equipment, where the sensors and circuitry are cooled to reduce noise and amplify the faintest detail. These systems are expensive but, as with any hobby or interest, you can put in as much or as little as you want; the results and reward are worth it.

• • •

If you want to develop your interest in stargazing or pursue the creative side of astrophotography, my advice is to take small steps and work up from binoculars and smartphones and small, capable cameras to more advanced DSLRs. Then you can invest in more advanced equipment if you still have the astrophotography bug. You will also get a great deal of experience while doing so, and you don't have to wear tweed or smoke a pipe to be good at it – just practise.

You can still carry on with your naked eye stargazing. There is no pressure, as stargazing is about you and what you want to take from it.

Chapter Fourteen

TO BOLDLY GO . . .

Well done, you've made it! Give yourself a massive pat on the back, especially if you always thought that stargazing was beyond you, or was something requiring books and equipment and studying. Stargazing is for everyone. Develop your interest as far as you like – the whole universe is now open to you.

There are huge benefits in looking up at the night sky. The simple act of stargazing boosts your wellbeing, helping you tune in to the world around and above you, embrace the stillness and absorb comfort from the mysteries of the night sky. You relax and wash away the stress of the day when the Plough presents itself to you, or you find your way around the Summer Hexagon without having to look at a star map or book. Ideally without getting cold, a stiff neck, or uncomfortable encounters with wildlife, or indeed humans, that could bring your stargazing session to an abrupt end.

Star hopping around the seasons and their associated constellations to easily locate and identify the prominent stars and the groups and collections of shapes they form is a magical experience. As is tracking the planets and the Moon through their annual cycle. And spotting a shooting star is still the highlight of my week. And again, we only need our eyes; no equipment is required.

The Secret World of Stargazing, as with my social media, tours and other activities, is written for those with a passing interest in astronomy or starting on their stargazing journey, to show them how easy it is to learn. I believe many of us look up after dark with feelings of awe and peace and amazement, and hopefully this book has made the night sky more understandable and accessible. And not only do I want it to be enjoyed by many, but I hope it also inspires you to relax, to feel a greater connection with the cosmos and to ask questions about what else is out there as you boldly go . . .

RESOURCES

Further Reading

Cohen, Andrew and Cox, Brian, *The Planets* (William Collins, 2019)

Cohen, Andrew and Cox, Brian, *Wonders of The Solar System* (William Collins, 2010)

Consolmagno, Guy and Davis, Dan M, *Turn Left at Orion – Hundreds of night sky objects to see in a home telescope and how to find them* (Cambridge University Press, 2011)

Dickinson, David and Cain, Fraser, *The Universe Today – Ultimate Guide to Viewing the Cosmos* (Page Street Publishing, 2018)

Dickinson, Terence and Dyer, Alan, *The Backyard Astronomers Guide*, (Fourth edition, Firefly Books, 2021)

Fry, Stephen, *Heroes – The myths of the Ancient Greek heroes retold* (Michael Joseph, 2018)

Fry, Stephen, *Mythos – The Greek myths retold* (Michael Joseph, 2017)

Graves, Robert, *The Greek Myths – The complete and definitive edition* (New edition, Penguin, 2017)

March, Jennifer, *The Penguin Book of Classical Myths* (Allen Lane, 2008)

Owens, Steve, *Stargazing for Dummies* (For Dummies, 2013)

Philip's Planisphere – For Use in Britain and Ireland, Northern Europe, Northern USA and Canada (Philip's, 2018)

Philip's Star Chart (Philip's, 2003)

Quicke, Greg, *Is the Moon Upside Down? – A Quicke guide to the cosmos* (Penguin Books Australia, 2020)

Sagan, Carl, *Cosmos – The story of cosmic evolution, science and civilisation* (Abacus, 1983)

Tirion, Wil, *The Cambridge Star Atlas* (Fourth edition, Cambridge University Press, 2011)

Apps and Software

Heavens-Above (Android)

Nightcap Camera (iOS)

Stellarium Planetarium (iOS, Mac, Windows)

Websites

www.meteorwatch.org (Stargazing news and events)

www.heavens-above.com (Satellite tracking and astronomy info)

www.space.com (Space news)

www.Universetoday.com (Space news)

ACKNOWLEDGEMENTS

As this is my first book, it has been a voyage of exploration in itself, and I would like to acknowledge my entire family for bearing with me and encouraging and supporting me while I knuckled down every day.

Most importantly, I would like to thank my partner Karen for all of her amazing support, and for being my rock and guide as I ventured forth into this new world.

I'd like to thank my elderly parents also for their wide-eyed enthusiasm and pride, which has driven me forward.

I mustn't forget my children for their help either – they have been perfect. As have the cats with their endless entertainment and company.

Thank you to all of those who have helped and inspired me on my stargazing journey.

ABOUT THE AUTHOR

Adrian West, better known as VirtualAstro on social media, is a writer, presenter and experienced independent astronomer with a lifelong passion for space, science, nature and the night sky.

Adrian has written astronomy and space-related articles for various online science websites, such as Meteorwatch and Universe Today, and guides and articles for the BBC, the Met Office, National Trust and other media outlets. He has also been a popular source for TV and radio news channels when something interesting is happening in the night sky.

He has a massive social media following, with one of the largest independent astronomy and space accounts on Twitter and Facebook. As well as his social media activity, Adrian often performs on stage in theatres with his popular and entertaining shows, including The Night Sky Show. In addition to this, he gives astronomy tours on a windswept hill in a dark sky location, showing groups of all ages the wonders of the night sky.

He's passionate about inspiring people to look up by being interesting, down-to-earth and fun. Adrian lives in southern England with his cats, partner and young family.

INDEX

Index

books to help you live a good life

Join the conversation and tell
us how you live a #goodlife

🐦 @yellowkitebooks
📘 YellowKiteBooks
📌 Yellow Kite Books
📷 YellowKiteBooks

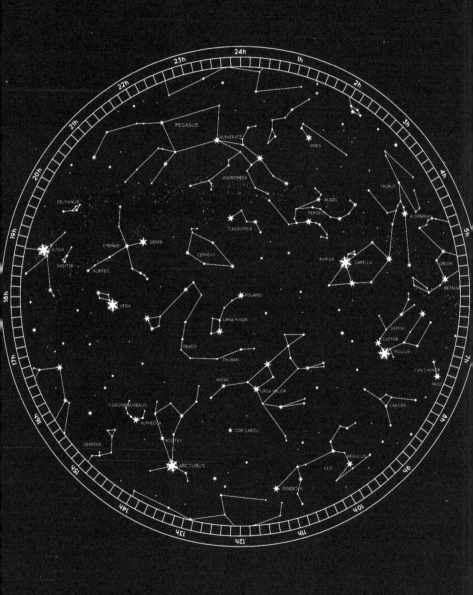

NORTHERN
HEMISPHERE